BIRD CONSERVATION
1

BIRD CONSERVATION
1

EDITED BY
Stanley A. Temple

Published for
THE INTERNATIONAL COUNCIL FOR BIRD PRESERVATION,
UNITED STATES SECTION
by
THE UNIVERSITY OF WISCONSIN PRESS

Published 1983

The University of Wisconsin Press
114 North Murray Street
Madison, Wisconsin 53715

The University of Wisconsin Press, Ltd.
1 Gower Street
London WC1E 6HA, England

Copyright © 1983
The Board of Regents of the University of Wisconsin System
All rights reserved

First printing

Printed in the United States of America

ISBN 0-299-08980-0, cloth
ISBN 0-299-08984-3, paper

Contents

Bird Conservation: An Annual Publication of ICBP STANLEY A. TEMPLE	vii
Restoration of the Peregrine Falcon in the Eastern United States JOHN H. BARCLAY and TOM J. CADE	3
The Bald Eagle in the Northern United States JAMES W. GRIER et al.	41
California Condor Reproduction, Past and Present NOEL F. R. SYNDER	67
The California Condor Recovery Program: An Overview JOHN C. OGDEN	87
Bird Conservation News and Updates	
Convention on International Trade in Endangered Species	103
Ramsar Convention	104
Fish and Wildlife Conservation Act	108
Federal Endangered Species Program	109
Restoration of Bald Eagles	109
Snake River Birds of Prey Area	113
Harris' Hawks Returned to Southern California	115
Peregrine Recovery Efforts	115
Status of Whooping Crane Conservation Efforts	117
Conservation Activities for Parrots	122
Status and Conservation of Woodpeckers	125
Tropical Deforestation and North American Migrant Birds	126
A Status Report on the Dusky Seaside Sparrow	128

Review of Bird Conservation Literature 133
  RICHARD C. BANKS, HELEN S. LAPHAM and
  L. RICHARD MEWALDT

Contributors 147

Bird Conservation: An Annual Publication of ICBP (U.S. Section)

Stanley A. Temple

WITH publication of this volume, the U.S. Section of ICBP initiates a new annual publication, *Bird Conservation*. The U.S. Section's purpose for existence is to promote the preservation of wild birds, especially in the Western Hemisphere and in relation to U.S. problems or U.S. activities abroad. *Bird Conservation* is one of the Section's means for achieving this purpose.

In recent years there has been much interest among professional biologists and concerned amateurs over the conservation of birds. Because birds are well-studied by ornithologists and because the public usually places high value on birds, conservation programs for them have been far more sophisticated and widespread than similar programs for other groups of animals. Progress in the field of bird conservation has been so rapid that exchange of information has often been inadequate. Furthermore, periodic, first hand reports on bird conservation programs rarely appear in print. Ornithological journals prefer not to publish the accounts because they are so "management oriented" and wildlife management journals prefer not to publish them because they are not final reports. Hence, the scientific community and the general public usually has to rely on second-hand reports in the popular literature to monitor progress of programs. Several of the most active bird conservation projects have, as a result, resorted to publishing their own reports. *The Peregrine Fund Newsletter, Brolga Bugle, Egg Rock Update, The Eyas*, and many others are examples of such attempts to keep the interested public informed.

Bird Conservation will function as a means of disseminating information on bird conservation activities. Each annual will include several major reports on specific conservation programs written by the biologists who are involved directly with the projects. Usually these reports will focus on a single or a few key issues. The first annual, for example, focuses on birds of prey with specific articles on conservation efforts for peregrine falcons, bald eagles, and California condors, three of the most important and often controversial conservation programs for this group. A second section of each annual will be "Bird Conservation News and Updates." In these shorter reports experts on a specific taxonomic group or type of conservation activity will summarize and review recent activities from the perspective of an involved and informed conservationist. Although these shorter reports will be primarily factual and reportorial in nature, critical comments and reviews will provide an added dimension.

The final regular section of *Bird Conservation* is a review of the recent literature dealing with bird conservation topics. Published accounts of the biology and conservation of endangered and threatened birds, population trends in declining species, and human activities that affect bird populations will be cited. In many cases, the citations will be annotated.

It is most appropriate that the U.S. Section of ICBP undertake this venture. The ICBP is the oldest international conservation organization, and its focus on birds has led to notable successes. ICBP accomplishes its work through its national sections in countries around the world. The U.S. Section is one of the largest and most active national sections, and bird conservation programs in the U.S. are among the most sophisticated and successful in the world. The U.S. Section is comprised of 19 constituent organizations that encompass the entire breadth of interest in bird conservation and that together have a combined membership of over 1.5 million. It is our hope that *Bird Conservation* will serve a valued function to those concerned enough with problems of bird conservation to have a desire to be well informed.

BIRD CONSERVATION
1

Restoration of the Peregrine Falcon in the Eastern United States

John H. Barclay and Tom J. Cade

THE original population of peregrine falcons (*Falco peregrinus*) in the eastern United States, approximately 350 breeding pairs (Hickey 1942), declined during the 1950s, and by the mid-1960s there were no known pairs remaining east of the Mississippi River (Hickey 1969). The extirpation of the eastern peregrines has been attributed to the biological concentration of persistent, organochlorine pesticide residues (DDT and related compounds) in the falcon's food chain and to the effects of these chemicals on eggshell formation and consequent reproductive failure (Peakall 1976).

In 1970 the Division of Biological Sciences and the Laboratory of Ornithology at Cornell University established a captive breeding program with the objectives of developing methods for raising peregrines on a large scale in captivity and eventually providing a source of falcons for introduction into the vacant, eastern breeding range (Cade 1974). The first 20 young peregrines were produced in 1973, and since 1976 more than 50 young have been raised each year (Cade and Fyfe 1978). Captive reproduction was large enough by 1975 so that experiments to develop effective methods for restocking peregrines could begin that year (Cade 1980).

In 1975, also, the U.S. Fish and Wildlife Service appointed an Eastern Peregrine Falcon Recovery Team, which has developed a Recovery Plan (Bollengier et al. 1979) that details the actions required to restore the peregrine in the East. The Recovery Plan is based on the concept of establishing a new population of peregrines by introducing captive-produced falcons into the wild, and its ultimate goal is the restoration of a self-maintaining population at a level of 50 percent of the number of breeding pairs estimated to have occurred in the 1940s or to a level the present environment will support. Since there are positive indications that DDT residues have declined significantly in the biota of eastern North America since the mid-1960s (Johnston 1974, Peakall 1976), released falcons should be capable of successful reproduction in the out-of-doors.

Releases have continued each year since 1975, and the Recovery Team set 1980 as the year for a full evaluation of the reintroduction program. Our report summarizes the first six years of effort to release peregrines in eastern environments and provides a basis for evaluating the success and effectiveness of the recovery program.

MATERIALS AND METHODS

Origin of Captive Breeding Stock

The original population of "Appalachian" peregrines or "duck hawks" indigenous to the eastern United States probably represented an ecotype that was best adapted to the primal conditions in the eastern deciduous and mixed deciduous-coniferous forests along major rivers and in mountains. Certain regional, physical characteristics (large size, black heads with little or no white in the cheeks) distinguished many of the individuals from falcons in other North American populations (White 1968). Unfortunately this population (gene pool) became extinct before any captive birds were brought under management for propagation, and consequently all captive-produced peregrines available for release in the East are the progeny of falcons from non-indigenous populations.

Participants at an Audubon conference on peregrine recovery in 1974 agreed that "the most promising ecologically-preadapted stock" should be used for release in the East (Clement 1974), and the Eastern Peregrine Recovery Plan (1979) further endorses the idea of releasing peregrines of different geographic origins and genetic makeups in order to identify the most suitable genomes for establishment in the eastern United States, namely those capable of producing peregrines that can survive and reproduce in the current environment.

The released birds were raised at the Behavioral Ecology Building, Cornell University, in a program operated by The Peregrine Fund, Inc.,

Table 1. Geographic origins of released peregrines

Parental origin	1975	1976	1977	1978	1979	1980	1981	Total
Nearctic tundra	10	15	14	14	9	27	30	119
Spain	0	1	8	7	6	8	7	37
Pacific N.W. coast	0	6	6	6	7	4	4	33
Mixed[a]	6	15	18	26	30	26	43	164
Total	16	37	46	53	52	65	84	353

[a] Includes offspring of parents from Alaska, California, northern Canada, Pacific Northwest coast, Australia, Chile, Scotland, and Spain.

and the Laboratory of Ornithology, or at companion facilities in Fort Collins, Colorado. Cade et al. (1976) and Cade and Fyfe (1978) have described the methods used to produce falcons in captivity.

The peregrines in the breeding colony at Cornell come from six main sources: The North American tundra (Alaska and Canada); the North American taiga (Alaska and Canada); the Pacific Northwest (Aleutians and Queen Charlotte Islands); western United States; Scotland; and Spain. Table 1 shows the numbers of peregrines derived from these and other sources that have been released in the eastern United States from 1975 through 1981.

Selection of Release Sites

Sites for the release of captive-produced peregrines were selected on the basis of both regional and local considerations. Initially sites were widely spaced with one in several eastern states. In order to increase the probability that released falcons would encounter one another for mating, we concentrated subsequent release sites in those areas where initial hacking success and survival of the released falcons had been highest. This strategy led to a pattern of clustered sites along the coast of New Jersey and in the Chesapeake Bay and a series of more widely scattered sites in inland regions.

The following requirements were considered in selecting the actual locations for release sites: The presence of suitable, open terrain where the falcons could effectively hunt; a sufficient quantity of prey; a minimum of human disturbance; and security from potential predators. These basic requirements could be met at two types of sites: Natural cliff sites that had historical records of use by peregrines and artificial sites, which consisted of city buildings or man-made towers that we placed in areas where breeding peregrines did not occur because of the absence of suitable nesting structures. All of the releases in the Chesapeake Bay and coastal New Jersey employed artificial sites, and all except one in the inland region involved natural sites on cliffs.

Peregrines most frequently nest on cliff faces, but they have been recorded nesting in a wide variety of other locations in different parts of their cosmopolitan range (*see* Hickey and Anderson 1969 for a review). It is sometimes assumed that peregrines nested exclusively on cliffs in the eastern United States, but a review of the literature shows that this was not the case since nesting had been recorded on city buildings (Groskin 1952) and in tree-top cavities in Tennessee (Ganier 1931). Some have argued that captive-produced peregrines should not be introduced into coastal marsh habitats because there are no historical precedents of peregrines nesting in these areas. There are two records that show that peregrines had, indeed, nested in coastal environments in the eastern United States. Jones (1946) found two broods of young peregrines in nests built by other birds, probably osprey (*Pandion haliaetus*) nests, in the Chesapeake Bay region of Virginia. He went on to say that peregrines had been resident and breeding in this area for at least 20 years. The other record is in the New Jersey State Geologist's Report (1890) which lists peregrines breeding in Cape May County at the southern tip of the state. This report does not state the circumstances in which the falcons were nesting so we can only speculate that they may have appropriated the nests of either bald eagles (*Haliaeetus leucocephalus*) or ospreys since this area of New Jersey contains no cliff formations.

Releases from specially constructed towers were initially designed to facilitate behavioral studies and to explore the possibility of modifying the peregrine's usual preferences for habitat and nest-sites. It was hypothesized that if peregrines were raised and released from such structures they would become "imprinted" to them and to the local area and that they would return to a similar structure and habitat for breeding (Cade 1980).

Release Technique

The technique used to release captive-reared peregrines is a modification of the falconer's practice of "hacking" (Michell 1900). Details of the procedures for hacking have been thoroughly described by Cade and Temple (1977) and by Sherrod and Cade (1978). Briefly they consist of placing broods of three to nine fledglings approximately four-weeks old in a protective enclosure (hack-box) at the release site. The falcons are fed in the box and released at approximately 40–45 days of age, when they are capable of sustained flight. Food is supplied at the hack-box for the next five weeks or until the young falcons become self-sufficient and disperse from the hacking locale, after which they are capable of leading an independent existence.

Marking Techniques

All released birds were banded with standard U.S. Fish and Wildlife Service leg bands. Since 1981 the leg bands of released birds have been anodized black to distinguish released birds from wild peregrines marked with normal bands. All of the birds released during 1975–77, about half of those during 1978–79, and one quarter of those during 1980, were fitted with radio transmitters during the two-three weeks after release. All the released birds during 1975–79 were banded with plastic colorbands of the type supplied by F. Prescott Ward (1976) for his International Peregrine Color Banding Program. About one half of the birds released during 1977–79 were marked with color-coded tail streamers made from one or two 25-cm strands of 1-mm diameter, non-conductive teflon wire-insulation. These streamers were attached to the proximal portion of the rachis of one central retrix. Some birds were experimentally marked with wing-tags in 1976, but their use was discontinued because they may have had a detrimental effect on the falcon's hunting abilities.

RESULTS AND DISCUSSION

Hacking Results

During 1975 to 1981, 81 releases involving 353 captive-produced peregrines (181 males, 172 females) were conducted at 36 locations in 9 eastern states and the District of Columbia (Figure 1). Twenty release sites were used for one year, 5 for 2 years, 2 for 3 years, 2 for 4 years, 5 for 5 years, and 2 for 6 years.

Table 2 summarizes the hacking results and the causes of mortality or other losses of released peregrines during hacking. Nine groups totaling 32 young were released by fostering to parentally motivated adults or by a combination of fostering and hacking. Although these birds were not by definition "hacked," they have been included in the totals of young released by hacking for ease of analysis. For the purpose of this analysis four weeks after release has been arbitrarily selected as the time when hacked peregrines are likely to be self-sufficient. During the third and

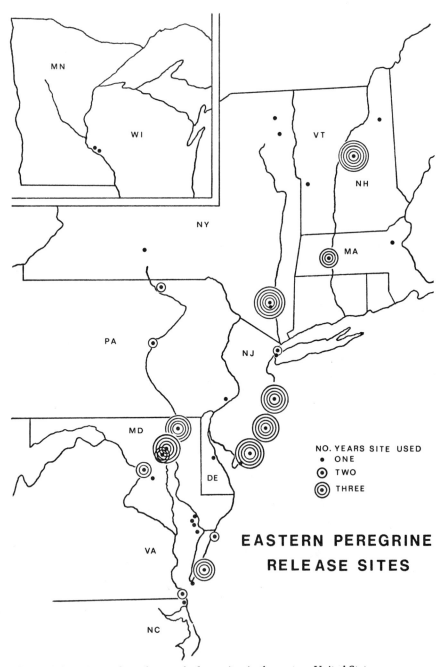

Figure 1. Location and yearly use of release sites in the eastern United States.

Table 2. Fate of hacked peregrine falcons up to four weeks after release

	1975	1976	1977	1978	1979	1980	1981	Total
Number peregrines hacked	16	37	46	53	52	65	84	269
Causes of mortality and other losses:								
Great horned owl predation	2	—	5	6	3	—	—	17
Retrapped due to owls	1	—	2	5	—	—	—	8[a]
Fox predation	—	1	—	1	—	—	1	3
Raccoon predation	—	2[b]	—	—	—	—	—	2
Injured by osprey	—	—	—	—	1[a]	—	—	1
Adult peregrine harassment	—	—	1	—	2	5	5	13
Died or returned to captivity	—	—	—	2	—	2	6	10
Forced away in storm	—	1	—	3	—	—	—	4
Electrocution	—	—	1[a]	—	—	—	1	2
Drowned	—	—	1	—	—	—	—	1
Fell into chimney	—	—	—	—	—	1	—	1
Flew into window	—	—	—	—	—	—	1	1
Drowned in air conditioning	—	—	—	—	—	—	1	1
Ensnared on building	—	—	—	—	—	—	1[a]	1
Disappeared prematurely	1	8	1	2	6	2	7	27
Total number lost	4	12	11	19	12	10	24	92
Number dispersed normally	12	25	35	34	40	55	60	261
Percent dispersed normally	75	68	76	64	77	85	71	74

[a]Returned to captivity.
[b]Occurred prior to release.

fourth weeks after release the falcons spend increasing amounts of time away from the release sites. Since most birds were not radio-tagged after two weeks following their release, it was difficult to verify mortality or to separate mortality from dispersal during the later stages of the hacking process. Some birds became independent as soon as two and one-half weeks after release, but they were exceptions. Other birds continued to return to the hack-site for food for six weeks or longer after release. The particular circumstances at each site and the behavior of the individual birds involved were taken into consideration in deciding whether disappearing birds had dispersed normally or not.

Mortality and other losses incurred during the hacking process ranged from 15 to 36 percent per year. The seven-year average was 26 percent, which yields a hacking "success rate" of 74 percent. Nine of the 92 birds lost during hacking died before they were released. These five instances have been included in the computation of the release results even though they occurred during the pre-release phase of the hacking process. Thus, of the total of 353 peregrines placed at sites for hacking, 344 were actually released.

The highest known cause of mortality was predation by great horned owls (*Bubo virginianus*). Eight birds were retrapped because of the imminent threat of predation at sites where more than one bird had already been killed by owls. Although these eight cases do not represent mortality, they should be included in the total of owl related losses because these eight birds were effectively removed from the population; therefore, the presence of great horned owls accounted for 25 (27 percent) of the 92 birds lost during hacking.

The 13 losses listed as "adult peregrine harassment" were actually cases of starvation that were precipitated by the presence of previously released subadult or adult peregrines. In 1977 one of a group of three birds was found starved seven days after it had been released. This bird was seen in the vicinity of the hack-site but was not observed to have returned to eat at the hack-box after being released. This individual, as were most of the birds intended for release, was fed by captive adults at the breeding facility prior to being placed at the hack-site. A subadult peregrine from the previous year's release was identified at the site on three occasions. Apparently the fledgling was attracted to it as a source of food. The subadult showed no overt signs of aggression but was obviously not parentally motivated to provide the young with food. Two birds in 1979, one at each of two sites, five of nine birds at one site in 1980 and five birds at two sites in 1981 were forced away from the hack sites by territorial subadult or adult falcons. In all of these cases the fledglings failed to return to the hacking station, and we therefore assume that they starved.

The 27 birds listed as "disappeared prematurely" in Table 2 represent cases in which birds disappeared at a point in the hacking process when it was unlikely that they could have been self-sufficient. Some of these birds disappeared under circumstances in which predation by great horned owls was suspected, but they have not been listed as such because no direct physical evidence could be found. The six birds which disappeared prematurely in 1979 included the entire group of four birds at one site. These birds all disappeared during the afternoon of the twelfth day after release. There were no indications of what might have happened to them, and it seemed unlikely that they could have been killed by a predator or have been brought down by accidents at the same time. Some information has since surfaced to suggest that the birds may have been intentionally killed or trapped, the only evidence, albeit circumstantial, that any of the released peregrines may have been the victims of human persecution during the hacking period.

Efficiency of the Hacking Process

The initial objective of the peregrine release program was to develop a reliable technique for establishing young, captive-produced peregrines in nature. The results from 1975 to 1981, during which 74 percent of the released birds survived to independence, indicate that the hacking process provides a reasonable degree of success. An appropriate measure of how efficient the hacking technique has been would be to compare mortality incurred during hacking with mortality of young "wild" peregrines during their post-fledging dependency period.

Estimates of first-year mortality in peregrine populations range from about 55 percent (Shor 1970) to 80 percent (Mebs 1960). These and other calculations of mortality rates do not indicate the seasonal distribution of first-year mortality, although most authorities agree that mortality is probably highest during the weeks immediately after the young become independent of their parents. Although quantitative estimates of mortality during the period from nest-departure to independence are not available for peregrines, a few studies of other raptor species have provided some data which suggest the magnitude of post-fledging mortality that might be expected in peregrine populations.

Hickey (1949) used band-recovery data to make life-table calculations of mortality in harrier (*Circus cyaneus*) populations. He estimated a 25 percent mortality of fledglings during the first three weeks after the young leave the nest. Harriers are similar to peregrines in some aspects of their population dynamics; they normally reach sexual maturity at two years of age, juvenile and adult mortality rates are comparable to those in peregrines, and the numbers of young raised per successful nesting attempt are only slightly higher than have been recorded in peregrine populations (Newton 1979). A 25 percent postfledging mortality rate might, therefore, be close to that occurring in wild peregrine populations during the period of dependency.

Estimates of postfledging mortality using band-recovery data must be considered as rough approximations because of the low probability of recovering banded birds during the short period involved. Burnham et al. (1974) reported the recovery of two fledgling peregrines near the eyries where they had been banded the previous year in Greenland, and they suggested that there may be a substantial mortality of fledglings before they become independent and leave the nesting areas.

Radio-telemetry has decided advantages as a means for determining postfledging mortality because unlike banding, most telemetered individuals can be accounted for whether dead or alive. Without telemetry most of the instances of mortality during hacking, and the causes, could not have been identified. There have been no large scale telemetry studies of wild, fledgling peregrines, but some recent research in the Snake River Birds of Prey Natural Area in Idaho has provided data on the extent of mortality in fledgling prairie falcons (*Falco mexicanus*).

Kochert (1976, 1977) calculated pre-dispersal mortality in prairie falcons of 22 percent in 1975, 12 percent in 1976, and 11 percent in 1977. In a nearby study area Peterson (1976) determined a 21–26 percent pre-dispersal mortality of the young from 14 eyries under observation. In these studies not all of the instances of mortality were verified through telemetry. Some of the fledglings were found dead during systematic searches of the nesting cliffs; therefore, these mortality rates must be considered minimum estimates because birds that died during the later stages of the pre-dispersal dependency period, when the young were flying greater distances from the eyries, would not have had an equal probability of being found. Given the inherent difficulties in arriving at accurate estimates of pre-dispersal mortality, the rates observed in these two other raptor species indicate that the 25 percent mortality incurred during hacking is probably within the range of mortality that might be expected for fledgling peregrines in wild populations.

The results during 1975 to 1981 demonstrated that the hacking procedure is an effective method for establishing young captive-reared peregrines in the wild. The birds released by hacking become self-sufficient hunters through a sequence of behavioral development that closely parallels the development of parented young in the wild. Comparative studies of the ontogeny of hunting behavior in hacked and wild peregrines indicate that the frequency of prey encounters may be higher for birds hacked in areas where prey is abundant (S. K. Sherrod, unpublished Ph.D. thesis). The two major causes of losses during hacking have been the premature disappearance, predation by great horned owls and harassment by adults. The premature dispersal of birds is an inherent problem with the hacking technique, but these types of losses can be minimized, at least during the first few days after release, by carefully timing the release of the falcons at an age when they will not be prone to take long-distance flights away from the hack-site. Harassment of young by returning adults and subadults increased in severity in 1980 and 1981. This problem has been difficult to avoid since most instances of harassment occurred at sites where there were no older birds present until after the young were released. Some harassment has even occurred at sites that had not been previously used for hacking. Most of the other causes of loss have been accidental and not the result of any apparent flaw in the hacking procedure. The one major shortcoming of hacking is the lack of adequate protection for the falcons against nocturnal predation by great horned owls.

Table 3. Hacking results according to type of site

	Tower sites		Urban sites		Cliff sites	
Year	Number of falcons hacked	Percent dispersed normally	Number of falcons hacked	Percent dispersed normally	Number of falcons hacked	Percent dispersed normally
1975	10	90	—	—	6	50
1976	14	71	—	—	23	65
1977	23	91	—	—	23	61
1978	29	76	—	—	24	50
1979	32	75	8	100	12	67
1980	33	94	20	85	12	58
1981	37	62	24	75	23	83
Total	178	79	52	83	123	63

Hacking Results at Natural and Artificial Sites

Table 3 summarizes the hacking results at natural and artificial sites. The percentage of birds successfully released at the artificial sites were consistently higher each year except 1981. The higher mortality at natural sites resulted primarily from a higher incidence of great horned owl predation. Owl predation occurred at 6 of the 13 cliff sites, and 20 (16 percent) of the 123 birds hacked at cliffs were lost to owls. Great horned owl predation occurred at 4 of the 17 tower sites, and only 5 (3 percent) of the 178 birds released at these sites were killed by owls.

In addition to being the highest known cause of mortality at natural sites great horned owl predation was the reason for the discontinuation of releases at five cliff locations. Planned releases in two regions, the Susquehanna and upper Mississippi River areas, were abandoned because of the threat of owl predation.

Great horned owls are strongly territorial and generally intolerant of other raptors in their nesting territories (Miller 1930, Smith 1970). As these owls have been known to kill young and adults of other raptor species (Wiley 1975), it is not surprising that they will readily kill young peregrines that do not have the protection of territorial parents. Owl predation has also occurred during the release of captive-produced peregrines in the western United States (Burnham 1978) and Canada (Fyfe et al. 1978). Great horned owl predation on wild peregrines has not been widely documented (Amadon *in* Hickey 1969:492–493), and the historical relationship between these two species in the eastern United States was not well known (Hickey 1942, Herbert and Herbert 1965). In Europe, the Palearctic equivalent of our great horned owl, the eagle owl (*Bubo bubo*) has been recorded preying on peregrines (Uttendorfer 1952 and Newton 1979). Terrasse and Terrasse (1969) reported that eagle owls apparently limited the breeding distribution of peregrines in one area of France, because it was not until after the eagle owl population disappeared that a small population of 17 nesting pairs of peregrines became established.

Even though great horned owls represent the greatest impediment to the successful release of peregrines at some historical nesting locations, their presence should not prevent the eventual reoccupation of these areas by adult peregrines. For example, previously released peregrines have been sighted at two of the cliff sites where releases were later discontinued because of owl predation. These two predators are sympatric throughout much of their ranges in North America, and they are no doubt adapted, probably through mutual avoidance of nesting territories, to coexist in close proximity.

Most of the instances of owl predation occurred during the first week after the falcons were released. It appeared that owl depredations could be prevented if resident owls were removed from the immediate vicinity of the hack-sites during the initial period after release when the falcons seemed to be most vulnerable. During 1975–78 a variety of raptor trapping devices were used to trap owls at several release sites. These measures were highly ineffective; only three owls were captured, because it appeared that the owls avoided bait animals and traps during the spring and summer when prey is abundant. Trapping efforts were intensified in 1979, and a tame "decoy" owl was used because it was thought that resident owls could be more effectively attracted by eliciting a territorial rather than a predatory response. This technique was effective, and 12 owls were removed from four release sites; however, falcons were later killed by owls at two of these sites. At one site six owls (four adults and two nestlings) were removed over a two-month period, but there were still two owls known to be in the area when trapping was discontinued. The trapping results in 1979 indicate that in areas where great horned owls are common, territories which are vacated as a result of trapping are likely to be quickly reoccupied by owls from adjacent territories.

Although the incidence of owl predation was reduced in 1979, the trapping of resident owls does not appear to be a feasible way to provide adequate protection for peregrines at hack. The results indicate that it will be costly in terms of mortality to release peregrines at many historical nesting sites using the hacking technique described here. The practical alternative is to concentrate release sites in areas devoid of great horned owls, such as coastal marshes, metropolitan areas, and perhaps in northern New England at high cliffs that are well removed from the mixture of agricultural and forest lands where these owls are common. Thus far no owl predation has occurred at the five cliff release sites in

Table 4. Recoveries of hacked peregrines

How obtained	Number of months between dispersal and recovery					
	0–3	3–6	6–9	9–12	12+	Total
Trapped by bander	16	—	—	—	2	18
Found injured—impact	8	1	—	—	—	9
Shot	6	1	—	—	—	7
Found dead—unknown	2	—	2	—	1	5
Found weak	3	—	—	—	1	4
Electrocuted	1	1	—	—	—	2
Poisoned	1	—	—	—	—	1
Total	39	3	2	0	4	48

northern New York and New England. The results at artificial sites, most of which were intentionally located in areas removed from owl habitat, and at the more northern cliff sites indicate that if owl depredations can be avoided then the hacking technique should obtain an 80–90 percent level of success.

Recoveries of Hacked Peregrines

Forty-eight of the 261 successfully dispersed peregrines have been recovered. Table 4 shows the methods of recovery and the time intervals between dispersal and recovery.

Dispersal

The release site and recovery location for the reported birds are plotted in Figure 2. Most of these movements involved latitudinal displacements, and there has been a general trend of movement in a southwesterly direction towards and along the coast. Only two of the recoveries occurred more than 500 km from the point of release, and there has been no consistent trend of longer movements with time. The two long-distance recoveries include a bird released in New Hampshire and found dead 12 months later in Saskatchewan, Canada, approximately 2800 km away and a bird released in Maryland that was trapped 2400 km away along the Gulf Coast of Mexico. Movements of this magnitude are not inconsistent with the extremes of movement that have been recorded for eastern peregrines in the past. Smiley and Smiley (1930) reported that a peregrine they had banded as a nestling in New York State was recovered three months later 2000 km away in central Nebraska. White

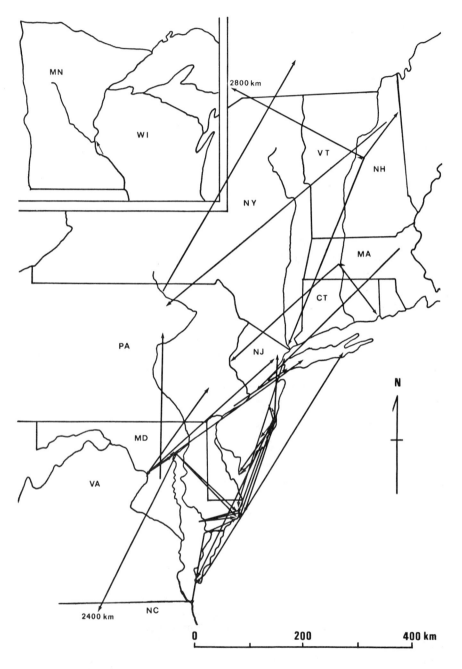

Figure 2. Dispersal movements of hacked peregrine falcons as indicated by release and recovery locations.

(1968) plotted the recovery locations of a sample of peregrines banded as nestlings at various locations in the eastern United States and has shown a pattern of dispersal for the former peregrines that is similar to the pattern in Figure 2.

Reported sightings of released peregrines have provided some additional information on fall and winter movements. Most of these sightings did not specify band numbers or tail-streamer colors, and so it has been difficult to determine the extent of movements since the release location of these birds is unknown. The greatest concentration of fall sightings has been in coastal areas from New York south to Chesapeake Bay. Winter sightings have occurred almost exclusively in coastal locations. Enderson (1965) and Bonney (1979) compiled the Audubon Christmas Bird Count records and have shown that peregrines wintering in the East tend to be concentrated in coastal areas. Some released birds, identified by tail-streamer color or color band, have been known to over-winter in the general vicinity of the sites where they were hacked in coastal New Jersey.

Three reports indicate the latitudinal extremes of movement of released birds. A wing-tagged peregrine seen in Key West, Florida, in January 1977, and a bird released in Maryland and trapped on the east coast of Mexico represent the most southerly known movements of any of the released birds. The most northern sighting was a peregrine with a tail-streamer seen near Cape Churchill, Manitoba, Canada, in June 1978.

Taken together, the information from recoveries and sightings indicates that the hacked peregrines are showing a pattern of dispersal that is typical of the "unoriented wandering" of first-year peregrines in sedentary populations (Hickey and Anderson 1969). Although there seems to have been a general trend of movement in a southerly direction towards coastal regions in the fall and winter, there is no evidence of migratory movement into Central and South America. It appears that the birds have been moving about in response to prey availability, which is highest in coastal areas during the late fall and winter, rather than in response to any strong migratory tendencies. These results stand in marked contrast to several reports from Central and South America of peregrines that have been released in Northern Alberta (R. Fyfe, pers. comm.).

Mortality After Departure from the Release Site

Band recovery data are one source of information that might provide some insight into the important question of how the survival of hacked peregrines compares with the survival of entirely wild peregrines. Unfortunately, it is difficult to arrive at firm conclusions, since the recovery data are limited. All of the recoveries have occurred within 14 months of dispersal, so it is impossible to derive life-table estimates of survival rates. The recovery data are, however, sufficient to enable some general comparisons with the recovery statistics of wild peregrine populations.

The temporal distribution of recoveries (Table 3), with a proportionally higher rate of recovery during the 0–3 month interval, is similar to the monthly pattern of peregrine recoveries that Enderson (1969a) reported.

Most discussions of mortality in peregrines point to the poor hunting ability of young birds as one of the major factors contributing to high juvenile mortality. Herbert and Herbert (1969) concurred with this idea, but they went on to conclude that:

> . . . by far the greatest factor in the mortality of juveniles is their unwariness where humans are concerned, and their consequent susceptibility to shooting and trapping in areas with heavily settled human populations.

The concept of releasing young peregrines which are reared in captivity not only raises the obvious question about how fit these birds will be for survival in the wild, but also about how vulnerable such birds might be to human persecution.

The hacking process has been designed to minimize human contact with the falcons and to prevent any associations between humans and food, in order to maximize "wildness" in the hacked birds. Because of the constant presence of humans (site attendants) and the lack of any negative reinforcement, the hacking process may, as an unavoidable consequence, produce a certain degree of tameness in the hacked birds that could increase their vulnerability to shooting.

Snyder and Snyder (1974) have discussed the possible effects of the taming of young raptors based on their studies of the nesting behavior of Cooper's hawks (*Accipiter cooperii*) in the southwestern United States. They reported that a significantly greater proportion of those nestlings which were familiarized with man, through repeated exposure to researchers, were later recovered as a result of "predation by man" than were those nestlings that received little or no exposure to humans, although the samples they compared are very small; they went on to conclude that the partial taming of young raptors may render them more vulnerable to shooting.

Through June, 1982, the recovery rate of hacked peregrines taken from the time of dispersal, has been 18 percent (48 of 261 birds dispersed). Shor (1970) examined over half of the records of peregrines banded in North America between 1924 and 1963 and determined a recovery rate of 12 percent (74 recoveries of 628 birds banded). Hagar (1969) reported a 16 percent recovery rate (N = 14) for the 88 peregrines banded as nestlings in Massachusetts between 1935 and 1947. Neither of these authors provided a quantitative analysis of the causes of recovery, except Shor stated that over one half were the result of shooting.

Shor (pers. comm.) pointed out that very few of the 74 recoveries he analyzed were made by banders capturing banded birds. Trapping efforts, especially for peregrines, have intensified over the last ten years in the eastern United States (Rice *in* Hickey 1969:279–280, Ward and Berry 1972, Ward 1976) so it follows that released peregrines would have had a greater likelihood of being recovered by banders than wild peregrines would have had in the past. In order to make comparisons of the recovery rates of released versus wild peregrines more commensurate it seems appropriate, therefore, to exclude recoveries by banders. Including these recoveries tends to inflate the overall recovery rate and reduces the frequency of recovery by other means, e.g. shooting. Excluding recoveries by banders (N = 18) brings the recovery rate of released birds down to 11 percent and the proportion owing to shooting to 23 percent.

Enderson (1969a) reported that 22 (50 percent) of 44 North American peregrines recovered as immatures were reported shot. In another sample of 58 recoveries of peregrines banded as nestlings in the eastern United States, 31 (53 percent) were reported shot, trapped, or poisoned (Enderson 1965, 1969c). Since Enderson was discussing the effects of human persecution we assume that those birds trapped were not caught by banders, but rather they were killed by trapping or held by falconers.

Considering, then, that hacked peregrines have been recovered at a rate comparable to that for wild peregrines during 1924–63 and that the frequency of recovery by shooting is less, it is tempting to conclude that peregrines released by hacking are no more vulnerable to shooting than entirely wild peregrines. Such a conclusion is valid, however, only if shooting pressure on peregrines has remained constant.

Newton (1979) compiled recovery data from Henny (1972) and Henny and Wight (1972) who suggest that there has been a decline in shooting pressure on raptors in the eastern United States in recent decades. The recovery rates of immature Cooper's hawks, American kestrels (*Falco sparverius*) and red-tailed hawks (*Buteo jamaicensis*) declined during 1940–60 as compared to the pre-1940 period. The percentages of recovered American kestrels and red-tailed hawks reported to have been shot declined markedly after 1940. These declines in the fractions of recoveries by shooting were presumably the result of protective legislation and/or a more tolerant attitude towards birds of prey. The possibility also exists that protective legislation and fines for killing raptors could have caused reductions in the reporting rates of shot birds without there having been a significant change in the incidence of shooting.

Shor (1970, 1976) analyzed the North American peregrine band-recovery data and discussed the effects of shooting pressure on recovery rates. He reported that only two percent of the peregrines banded in 1968 and three percent of those banded in 1969 had been recovered within five years. Shor attributed these declines in recovery rates to reductions in shooting pressure, and he went on to conclude that a recovery rate of four percent or less would be likely for peregrines in the 1970s. Since Shor did not indicate the circumstances under which the peregrines were banded in 1968 and 1969, we examined the records from the Bird Banding Laboratory. These records, which indicate only the location of banding, show that 68 percent of the 255 peregrines banded in these years were banded in Alaska or northern Canada, and most of the remaining records were from states where northern migrants are most frequently seen. In an earlier report Shor (1970) analyzed the recovery rates of different subpopulations of North American peregrines. These recovery rates ranged from 1.9 percent for birds banded as nestlings in Alaska to 23 percent for arctic migrants trapped in Texas during the fall. The recovery rate for resident peregrines from the United States and southern Canada was 12.1 percent, and for arctic migrants banded on Atlantic beaches it was 6.3 percent. Shor attributed these differences in recovery rates to regional differences in shooting pressure. For instance, arctic migrants are known to move through Atlantic coastal areas quickly during the fall (Ward and Berry 1972); therefore, they are exposed to shooting for less time than resident peregrines.

Shor's (1976) hypothesis that the recovery rate for peregrines in the 1970s should be on the order of 4 percent seems to reflect a shift in the emphasis on where and when peregrines have been banded more than any significant reduction in shooting pressure on the species. The recovery rate for the species would be expected to decline if proportionally more birds have been banded in recent years from those subpopulations with low recovery rates; therefore, the fact that released peregrines have been recovered at a rate higher than Shor expected for the species and the fraction of recoveries by shooting is also higher than he expected should not be taken as an indication that hacked peregrines are more vulnerable to shooting than wild peregrines. There would need to be contemporary recovery data from a resident population of peregrines in the East in order to draw any conclusions regarding the vulnerability of hacked peregrines to shooting. However, based on the existing recov-

Table 5. Returns of hacked peregrines to release sites

Region	Year returned						Total
	1976	1977	1978	1979	1980	1981	
Coastal New Jersey	2	2	5	8[a]	10	11	38
Chesapeake Bay	1	0	2	2	2	11	18
Inland	2	3	3	4	2	2	16
Total	5	5	10	14	14	24	72

[a]Includes a bird which was released in the Chesapeake Bay region that returned to a site in coastal New Jersey.

ery data, it can be concluded that the peregrines released by hacking have been subjected to no more shooting pressure than the wild, resident peregrines were prior to their decline.

Returns of Hacked Peregrines to Release Sites

The yearly totals of peregrines which returned to release sites in each of the three release regions are summarized in Table 5. Most, but not all, of the birds considered to represent returns were positively identified as released peregrines. In order to exclude possible sightings of migrant peregrines, these returns only include sightings of peregrines at or near release sites during the late spring or summer. The total of 72 returns does not represent 72 different individuals because approximately 15 birds have returned for two or more consecutive years. Taken together these birds account for about 35 of the 72 returns listed.

Except as noted in Table 5, all the birds considered to represent returns have come back to the region where they were released. This trend is consistent with the phenomenon of regional philopatry in wild populations of peregrines (White 1968) and other species (Newton 1979).

The return data from Table 5 are used to calculate return rates for each region in Table 6. The totals used to calculate the rates for the coastal New Jersey and the Chesapeake Bay region have been adjusted from the totals in Table 4. Those birds which returned for two consecutive years are taken to represent one return in Table 6. Also, 2 birds that were released in the Chesapeake Bay region but which returned to sites in New Jersey are included in the Chesapeake Bay total since these rates are based on the number of birds released within each region. The 1981 release totals are not included in these calculations because they have not yet had the opportunity to return.

Table 6. Regional return rates of hacked peregrines, 1976–1981

Region	Number of falcons hacked	Number of falcons dispersed[a]	Number of returns[b]	Number hacked per return	Number dispersed per return
Coastal New Jersey	51	41	19	2.7	2.2
Chesapeake Bay	62	51	6	10.3	8.5
Inland	91	54	14	6.5	3.9

[a] 1981 results not included.
[b] Individuals which returned in conservative years are counted as one return.

Discussion of Return Rates

Although peregrines ordinarily do not become sexually mature until two years of age, first-year birds have been known to occupy territories and engage in breeding but usually at inferior eyries or when vacancies exist at the traditionally occupied superior eyries and adult replacements are few (Hagar 1969, Ratcliffe 1980). In a normal population the presence of territorially dominant older birds no doubt restricts the opportunities for subadults to acquire nesting territories. The population of released peregrines is, however, demographically unique because it consists entirely of young age-class birds. Owing to the absence of any competitive exclusions of subadults by older breeding birds it could be expected that returning first-year birds would tend to occupy the most suitable nesting sites available. The average return rates are, therefore, the best measure of the suitability of the release sites in each region for eventual occupancy by breeding pairs.

Breeding peregrines were historically rare in coastal New Jersey and the Chesapeake Bay region because of the lack of suitable nesting structures. By releasing peregrines from artificial eyries in these regions we expected that the birds would return to them to breed, because the hack towers are superior to any other locally available alternative nesting structures. The locally available food resource is, however, an equally important requisite of a peregrine's nesting environment.

The disparity between the return rates for coastal New Jersey and the Chesapeake Bay area is best explained in relation to prey availability in the immediate vicinity of the release sites in these two regions. The hack sites in New Jersey are situated in coastal salt-marshes where prey is both more abundant and more readily available to peregrines than in the vicinity of those sites in the forested areas of the upper Chesapeake Bay. In addition to the high return rate in New Jersey, the birds returning there have consistently shown the greatest affinity for the hack-towers.

It is interesting to note that the return rate for coastal New Jersey has increased numerically, i.e. more birds dispersed per return, since a previous analysis based on releases during 1975–79 (Barclay 1980). Established pairs at the New Jersey towers during 1980 and 1981 have effectively reduced the occurrence of subadult returns at these towers and thereby produced a numerically higher (poorer) rate of return.

Beginning in 1978, new release sites in the Chesapeake Bay region were established in the lower bay region in coastal marshes that are ecologically similar to the coastal marshes in New Jersey. Since this shift to releases in the lower bay marshes, the rate of return and occupancy of these towers has been comparable to that experienced at the coastal New Jersey towers.

The only adult which has established a territory in the upper Chesapeake Bay region did so on a building in Baltimore, Maryland, approximately 15 km from the site where she was released. As far as is known, this bird has never returned to her release site, probably because the metropolitan environment she is occupying is superior in terms of available prey, feral pigeons. The total absence of territorial adults at any of the towers in the forested areas of the upper Chesapeake Bay suggests that artificial sites located in areas of marginal prey availability may fall below the threshold of acceptability as nesting sites, or that peregrines avoid towers located in habitat occupied by great horned owls.

Another more subtle factor that could be contributing to the higher rates of return to sites in coastal marsh environments may be related to the conspicuousness of the hack-towers. In his discussion of the psychological aspects of bird distribution, Lack (1937) suggested that birds instinctively seek the visually prominent features of a habitat. Hack-towers in salt marshes are, indeed, visually prominent in these otherwise relatively featureless landscapes. In addition to their possible "psychological" attractiveness, a prominent structure like a hack-tower may have a more immediate value to a peregrine as a hunting station. It is interesting to recall that in Hickey's (1942) discussion of the ecological magnetism of traditional nesting cliffs, he stressed the relationship between the attractiveness of cliffs to breeding peregrines and the degree to which they dominate the surrounding countryside.

The return rate at inland sites may not be strictly comparable to the rates observed in the two regions where artificial sites have been used. One fundamental difference has to do with the ease of detectability of peregrines on towers as opposed to cliff faces. Since the inland sites have been geographically more widely distributed (Figure 1), birds returning to the general vicinity of the sites where they were released might have gone undetected because there were no nearby release sites and no observers. In addition, cliff sites have been located in areas that

Table 7. Estimated survival of peregrines released by hacking[a]

Cohort	1975	1976	1977	1978	1979	1980	1981	1982
1975 cohort	(16)[b] 12[c]	5	4	3	2	1	0	0
1976 cohort		(37)25	11	9	7	6	5	4
1977 cohort			(46)35	16	13	10	8	6
1978 cohort				(53)34	15	12	10	8
1979 cohort					(52)40	18	14	11
1980 cohort						(65)55	25	20
1981 cohort							(84)60	27
Total sub-adults and adults		5	15	28	36	47	62	76

[a] Assumes 56% mortality from dispersal to one year of age and 20% yearly mortality thereafter.
[b] Number of birds released.
[c] Number of birds dispersed.

contain many alternate nesting sites, i.e., historical eyries. The objective of releasing peregrines in these areas is not necessarily to establish birds at the cliffs used for hacking, but rather to establish them at any of the alternative sites, some of which might be superior breeding sites in terms of local prey availability or other more subtle requirements. For these reasons the return rate in the inland region, where only about one out of every four birds dispersed has been resighted, may not be truly indicative of the effectiveness of releases at cliff sites.

The results at cliff sites show that if the initial problem of high hacking mortality can be overcome, then the frequency of return should augur well for eventual pair formation. This should be possible by confining cliff releases to higher cliffs in northern New York and New England.

Population Considerations

Estimate of Population Growth

Table 7 contains an estimate of the growth of the population of released peregrines. This estimate assumes that the survival rates of hacked peregrines are comparable to those of wild peregrines, and the calculations are based on mortality rates of 66.7 percent for first-year birds and 20 percent for subadults and adults (Young 1969). First-year mortality is usually applied to a given cohort from the time of nest-departure up to the following spring. It has already been established that there has been an average of 26 percent mortality of the hacked birds from release to independence (Table 2). Therefore, by applying a 55 percent mortality to the birds dispersed each year the result is an overall first-year mortality (from release to one year of age) of 66.7 percent.

As an indication that a 55 percent mortality after departure from the release site may be reasonably accurate, it can be pointed out that, of the 12 birds that dispersed in 1975, there were 5 known to be alive in 1976 (Table 5), which compares favorably with the five that theoretically should be alive according to Table 7. Also, the overall return rate for the coastal New Jersey region of 21 returns out of 52 birds dispersed (Table 6) is near the theoretical maximum that could be expected given a 50–60 percent post-dispersal mortality. The combined returns for all regions for the years 1977–81 (Table 5) have averaged about 33 percent of the numbers of birds that should be alive according to the figures in Table 7. These results should not necessarily be taken as evidence of higher mortality during these years. Simply because a given number of birds that theoretically should be alive do not return to release sites by no means indicates that they are no longer alive. Some of the factors possibly influencing return rates and detectability of returning falcons have already been discussed. Furthermore, it could be expected that a certain percentage of first-year birds would not return to release sites because within the labile context of dispersal behavior some individuals are probably more predisposed to settle away from their natal areas than others (White 1968).

The population growth depicted in Table 7 should be considered as a rough approximation at best. Chance events are important factors in the dynamics of natural populations, especially in the growth of small founding populations (MacArthur 1972), and mortality is rarely as constant as depicted in this simple population model; therefore, any random fluctuations in mortality can have proportionally large effects on the growth of the small founding population of released peregrines. Nevertheless, the population estimates that this model provides are useful simply to demonstrate the given the limitations of a high first-year mortality, large numbers of birds need to be released in order to reestablish a sizeable population.

Population Projections

The goal of the Eastern Peregrine Recovery Plan (1979) is the reestablishment of a population with one half the number of breeding pairs that formerly existed in the East or to a level the present environment will support. Since Hickey (1942) estimated approximately 350 breeding pairs of peregrines in the eastern U.S., the recovery goal would be 175.

Cade and Temple (1975) addressed the question of how many peregrines would need to be released and how many years it would take to approach the Recovery Plan goal. They developed a deterministic model that incorporated mortality and breeding success statistics for a stable population; breeding at three years of age, 50 percent breeding success, 2.0 young raised per successful pair, 66.6 percent first-year mortality and 20 percent yearly mortality thereafter. Their model was based on yearly releases of 250 young, and their calculations show that after 15 years of releases involving 3,750 birds there theoretically should be 146 successful breeding pairs in a total population of 1,180 birds; however, their calculation contains an error which produces a faster rate of growth than would actually be the case. The Eastern Peregrine Falcon Recovery Plan (1979) includes a similar projection based on the same population parameters except breeding at two years of age. These calculations are based on an initial release of 100 young and a 10 bird per year increase up to 240 young released in the fifteenth year, and they show that there should be 92 successful breeding pairs after 15 years of releases totaling 2,440 young. Both of these projections clearly demonstrate that large numbers of young will need to be released in order to reestablish sizeable populations of breeding pairs; however, both are based on rather optimistic figures for the numbers of young expected to be available for release in the years ahead.

Figure 3 shows a population project based on the numbers of young released during 1975–81 and a conservative estimate of a 10 bird per year increase up to a maximum of 100 young released per year during 1984–90. This projection was made using a stochastic computer program that was developed by J. W. Grier (1976) for modeling the growth of a population being introduced into an area by the yearly release of young. Because this model incorporates elements of random chance for selected reproductive and mortality rates, it provides a somewhat more realistic projection of population growth than can be obtained using deterministic models which assume constant natality and mortality. Since each simulation would theoretically yield different results, five simulations were performed in order to obtain a representative range of possible population sizes. The mortality and breeding success statistics used for each simulation are taken from Young (1969), and they are based on the dynamics of a stable population. The results in Figure 2 show that after 16 years of releases totaling 1,205 young there could be from 152 to 203 breeding-age birds in a total population of 294–375 birds. Assuming a 60 percent level of breeding success, there would be 45–60 successful breeding pairs in the spring of 1991.

It should be noted that the population growth depicted here is depen-

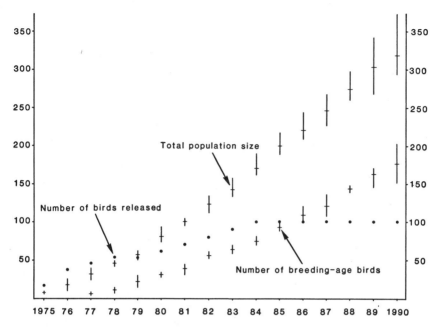

Figure 3. Projected growth of the population of released peregrines. The values plotted are the combined results of five computer simulations that were performed using a stochastic model for population growth and they represent the range and mean for the number of birds alive at the end of the calendar year, after reproduction and mortality for that year has taken place. These simulations were run using the following parameters; sexual maturity at two years of age, 60 percent of breeding age females successful, 2.5 young raised per successful breeding female, 66.7 percent first-year mortality, 20 percent subadult and adult mortality.

dent upon the annual release of captive-produced young. According to this model if releases were discontinued, population growth would cease and the population would stabilize because stable parameters were employed. In reality it would be expected that the population would gradually increase, as long as there were no problems with reproductive performance. The subpopulations in the Chesapeake Bay and Coastal New Jersey areas would level off at their respective carrying capacities which are determined by the number of nesting towers. The interim goal of the recovery effort is to establish 30 nesting sites along the Atlantic Coast and in the Chesapeake Bay. This number of sites is considered to be the minimum to support a small self-maintaining population, and the excess production of young from this population might then disperse into adjacent areas to repopulate some of the historical cliff eyries at inland locations.

Reproductive Behavior

In 1976 a male returned to a tower site in New Jersey (Sedge Island) and established a territory at that site where he had been hacked in 1975. This same bird returned in 1977, and being sexually mature, he reacted parentally to the hacked young that year by feeding them from his kills, which he brought regularly to the tower. He adopted the brood and provided nearly half of their food until they became independent. In 1978 an adult male, which may have been a different bird, returned to this same site and also adopted the brood of hacked young. A subadult female returned later in the season and formed a pair-bond with the male, but she did not act as a parent to the young; rather, she expressed some aggressive behavior. This instance represents the first known pairing by any of the peregrines released in the East.

In the spring of 1979 an adult male and subadult female were paired at another tower-site in New Jersey near Manahawkin. The female laid at least two eggs in May. During incubation there were numerous aggressive encounters between the falcons and crows (*Corvus ossifragus* and *C. brachyrhynchos*). On two occasions a crow flew inside the hack-box to seek refuge from the attacking male, and both times the female killed the crow inside the box. Shortly after the second incident, the falcons' incubation attentiveness declined markedly. The nest was inspected, and the eggs were no longer present.

A fractured egg was found at the base of the tower. Analysis of this egg indicated normal shell thickness compared to pre-DDT standards. This finding was especially significant because it was the first evidence that peregrines residing in coastal areas were not accumulating levels of pesticide residues that might inhibit reproductive success. Concerns had been expressed that peregrines in coastal areas might be exposed to high levels of pesticide residues if they preyed heavily on migratory, semi-aquatic birds (Peakall 1976). Pairs of first-year peregrines were present at the other two coastal New Jersey towers in 1979, but neither attempted to nest.

In 1980, three pairs of peregrines were again present at the coastal New Jersey towers. Two pairs laid eggs and produced young, three at one site and one at the other. One captive-produced chick was added to each of these broods and all of the young fledged. The third pair, consisting of an adult male and subadult female did not nest although nest-scraping, courtship and copulations occurred. They encountered interference from a pair of barn owls (*Tyto alba*) that attempted to nest inside the hack-box.

Other nesting activity in 1980 included two pairs of peregrines which raised two young each at cliff sites in eastern Maine and southern Quebec. The female of the pair in Quebec was verified by color band as a bird released in the United States in 1975 or 1976. The male of that pair was not banded and is presumed to be a "wild" bird, perhaps of taiga origin. The origin of the pair in Maine is unknown.

In 1981 four pairs of peregrines were present at the coastal New Jersey towers. An additional tower had been constructed in 1980 and it was immediately occupied by a subadult female that year. Three pairs hatched a total of eight young. The three young at one site were eaten by a raccoon (*Procyon lotor*) that managed to climb over the predator guards on the tower. The fourth pair did not nest since a pair of barn owls again gained access to the hack-box and thus preempted the nest site. Two additional pairs were present at coastal towers in the Chesapeake Bay region in 1981, but neither attempted to nest since one member of each pair was a subadult.

Also in 1981 a pair of Peregrines nested and raised two young at a historical cliff eyrie in the White Mountains of New Hampshire. The female of this pair was banded and is believed to be a bird released 25 km away at a cliff site that has been used for releases since 1976. The origin of the male of this pair is unknown, since it was not determined whether he was banded. The pairs in Quebec and Maine were absent in 1981 at the cliffs where they were observed in 1980.

In summary, there have been six successful nestings of released Peregrines in the eastern United States. A total of 14 young have been hatched, four by two pairs in 1980 and ten by four pairs in 1981.

In January of 1978 a female (Scarlett) released at Carroll Island, Maryland, in 1977 took up residence on the tallest building in Baltimore, the U.S. Fidelity and Guaranty home office, where she has remained ever since. We provided two gravel-filled nest-boxes on the 33rd-floor ledge, and in April of 1979 the unmated female laid three eggs. Young peregrines were later substituted, and the female raised all four to independence. Two adult males were released at separate times in 1979 in an attempt to provide a mate, but neither bird stayed in the area. In 1980 the female was still unpaired. We removed her first clutch of eggs and released a two-year-old male. The two birds quickly formed a pair-bond, but the timing was such that the male could not have fertilized the second clutch of eggs. Young were provided again, and both adults participated in rearing them to independence. The male was found dead in Baltimore in the fall of 1980. A necropsy indicated that he had died of strychnine poisoning. The female remained unpaired in the spring of

1981 and again laid four infertile eggs, as we had no suitable males for release as potential mates. Five captive-reared young were provided, and she again reared them.

Analyses of 14 infertile eggs laid by Scarlett in Baltimore in 1979, 1980, and 1981 reveal DDE residues averaging only 4.4 (range 3.1 to 5.7) ppm wet weight and eggshells ranging from 0 to 14 percent thinner than normal (Ratcliffe Indices ranging from 1.59 to 1.98). Most encouragingly, DDE residues have decreased and eggshell thickness has increased progressively over the three years. From an average of 5.3 ppm and a thickness index of 1.68 in 1979 (N=3) to 3.4 ppm and thickness index of 1.88 in 1981 (N=4). An addled egg collected at one of the New Jersey nests in 1981 contained 6.5 ppm wet weight of DDE, less than 0.5 ppm of five other chlorinated hydrocarbon compounds, and 10.0 ppm PCB, with a shell thickness index of 1.89. Another addled egg from New Jersey in 1980 had a thickness index of 1.88. These results indicate low levels of contamination, and the values are below the levels that Peakall (1976) considered to be critical to the hatchability of peregrine eggs. Data were provided by the U.S. Fish and Wildlife Service, Patuxent Wildlife Research Center, through the courtesy of Dr. Oliver Pattee.

CONCLUSION

After a lapse of 25 years, peregrine falcons are again nesting in the eastern United States. They are not descendants of the "duck hawks" that falcon enthusiasts once knew, for that population of peregrines was totally extripated by DDT poisoning. Nor are they nesting on the rocky crags overlooking the Connecticut, Hudson, and Susquehanna Rivers as the duck hawks once did. Instead, they are nesting on special towers in coastal salt marshes and on buildings in cities. They are, nonetheless, bona fide peregrines of the species *Falco peregrinus*, produced in captivity by parents taken from various geographic populations extraneous to the eastern United States.

From its beginning in 1970, everyone involved in the eastern peregrine recovery effort has clearly understood that restoration of a nesting population of peregrines in the East would depend upon the introduction of falcons drawn from non-indigenous sources. The issues involved were first thoroughly discussed at a conference sponsored by the National Audubon Society (Clement 1974). They were further reviewed in great detail by the Eastern Peregrine Falcon Recovery Team in preparing the official recovery plan for the U.S. Fish and Wildlife Service (Bollengier et al. 1979), and the Office of Endangered Species conducted its own internal review of the matter in 1977, following the President's Executive Order on the introduction of exotic organisms (E.O. 11987, 24 May 1977). In 1977, the Fish and Wildlife Service issued an official policy statement supporting the use of non-indigenous peregrines for release in the eastern U.S., a policy that is still in force today.

Since the propriety of releasing non-indigenous or exotic falcons into the "natural ecosystem" of the eastern states continues to be questioned by some people, and since the A.O.U. committee on resolutions specifically addressed the subject in a resolution passed at its 1979 annual meeting, we would like to recapitulate, once more, the salient and, we believe, justifiable reasons for the course of action undertaken. Our reasons are based on the following views about the conservation of biological diversity.

First, the Endangered Species Act rightly focuses on the species as the taxon of fundamental importance for preservation. Species are populations the members of which freely interbreed and exchange genes in nature, or that have the potential for doing so, but that are reproductively isolated from all other populations of similar organisms. Reproductive isolation is the key to the definition of species and also to the scope of the need for preservation, because all members of a species, regardless of how widespread or how many populations have been described by taxonomists as "subspecies," share a closed, cohesive, coadapted gene pool and a common epigenetic system of phenotypic development (Mayr 1963, Selander 1971). To be sure, a wide-ranging species such as the peregrine falcon is likely to consist of a number of "ecotypes" that are to some degree specially adapted to local environments, but if the work of Corbin (1977) and others who have been studying "genetic distance" among avian taxa applies to falcons, then probably as much as 90 percent of the total genetic diversity of the peregrine falcon is shared in common by all local breeding populations. This estimate of similarity led Corbin (*op. cit.*) to conclude: "It is likely that coadapted gene pools of birds differ somewhat among geographic areas. However, the genetic identity data suggest that for management purposes the origin of individuals being used to repopulate areas following local extinctions need not be a major concern of the program."

Second, the primary purpose of work under the Endangered Species Act is to restore species, as far as possible, to their original density and distribution prior to their endangerment; or, since that goal is usually not possible in today's world, at least to secure a free-living population in some suitable range. The purpose is not necessarily or only to preserve local genomes but to maintain the species as a population of ecologically adapted individuals and to increase their numbers in geographic regions where they have declined or disappeared.

Approximately 2.5 million square kilometers of former nesting range for peregrines now lie vacant east of the Rocky Mountains. As far as we know, the original breeding population has been totally extirpated for about 20 years. Any peregrines that become re-established as breeders within this vast range will be in some sense "exotic" or non-indigenous, whether they come in naturally or by human intervention.

The chances for natural re-establishment on this range are remote. The Great Plains constitute a partial barrier against dispersal from a sparse, poorly reproducing population in the western states, and northern breeders show little proclivity to settle and nest in mid-latitude habitats over which they migrate, although the recent nestings in Quebec and Maine indicate some possibility for a southward expansion of falcons from northern breeding locales. Thus, the situation is totally unlike that in Great Britain, where a resident population reduced to about 44 percent of its original numbers has shown a dramatic recovery since 1963 (Ratcliffe 1980).

There are several ways to go about establishing a new population. We can try to discover the existing peregrine population with the greatest number of genetic similarities to the extinct duck hawks and use that stock for our introductions, in the expectation that such falcons will have the maximum possible fitness for establishment in the East. Based on geographic proximity (since we know nothing about the actual gene pools), peregrines from Labrador might be the closest ones; based on morphological similarities, those from northern Alberta might be. Unfortunately, all such potential candidates are themselves severely endangered, and the removal of individuals from these populations for use in an eastern recovery program would meet strong resistance.

We can look for peregrines that are the closest ecological counterparts of the former eastern birds and try them. None of the existing North American populations really qualifies as well as, say, French or German peregrines do. But they are exotics, and moreover they are also severely endangered.

A third possibility put forth by William Drury (1974) and Ian Nisbet (*see* Clement 1974) represents a refreshing counter to the objections often raised about "exotics" and "mongrelization" of races. They suggest that breeders of captive peregrines should deliberately mix their stocks to achieve the greatest possible degree of genetic variability in the genomes and then to release these "hybridized" individuals into the vacant breeding range and let natural selection pick those individuals that are fit for the present environment of the East. Such a procedure recognizes our inability to determine a priori which kinds of peregrines are adapted for survival and reproduction in the current eastern environment and also comprehends the severe problems of inbreeding that often occur in small populations of animals, captive and wild (Ralls et al. 1979). It is worth calling attention here to the elaborate ways in which birds have evolved social behaviors that prevent inbreeding (*see* Koenig and Pitelka 1979).

Following Drury's lead, our basic working assumption has been that if we can release enough individual peregrines with some degree of fitness for the eastern environment, then natural selection will have a chance to work toward increasing fitness, so that after several generations a well adapted population of peregrines will emerge. If the same selective forces that produced the original duck hawks are still operative, then it is reasonable to predict that the new population will converge genetically and phenotypically on the old; if, as seems more likely, new and different selective forces are now associated with the much altered eastern environment, then a somewhat different peregrine will result. The difference will not be noticeable to 99.9 percent of the people who watch peregrines, and the new population will still belong to the closed gene pool that is represented by individuals of the species *Falco peregrinus*.

Since 1975 we have released 353 peregrines in the eastern United States. They represent genomes derived from breeding populations in Spain, Scotland, Chile, the Canadian tundra and taiga, the Alaska tundra and taiga, the Aleutian Islands, the Queen Charlotte Islands, and California. Many of these released peregrines have not only managed to survive in the eastern states but have also returned as adults to the environs where they were first established. Eight pairs have formed, two have successfully produced young of their own for two years, one for one year, and a fourth has reared fostered young. In addition one female has paired with a wild male at an eyrie in southern Quebec, and she produced two young in 1980.

The genetic backgrounds of these successfully established birds are as follows. The female in Baltimore is from a California male and a Chilean female; she was mated with a male of Nearctic tundra-taiga origins in 1980. The breeding pair at Brigantine, New Jersey, consists of a female derived from the Queen Charlotte Islands (a Peale's falcon) and another male of California-Chilean extraction. The Manahawkin, New Jersey, breeding pair consists of an Alaskan tundra-taiga male and a female of mixed Alaskan tundra-Queen Charlotte Island parentage. The Sedge Island male is Spanish, the female, Nearctic tundra. The female in Quebec is either of tundra origin or a tundra-taiga mixture; she has not been individually identified.

Despite their diverse genetic backgrounds, these successfully established peregrines have converged remarkably close in their main biological habits toward the former duck hawks. First of all, they are not highly migratory—some are quite sedentary—despite the fact that many of them derive from populations that are highly migratory in their natural ranges. By contrast, the Canadians, who have been releasing peregrines in northern Alberta and other northern locales, have had several of their birds reported from as far away as Mexico City and Belize south into northern South America (R. Fyfe, *in litt.*), exactly where one would expect them to go, following the natural migratory habits of the populations into which they had been introduced.

So far, based on ten nestings by five females, the timing of reproduction corresponds closely to the late March-April period of egg-laying that characterized the breeding season of the former duck hawks. Response to photoperiod and other seasonal timers of reproduction does not seem to be precisely influenced by genetic background, and earlier concerns about the complexity of the natural photoperiods experienced by the Arctic, migratory peregrines, for example, have proved to be groundless.

The released peregrines have adopted trophic relations virtually identical to those of the old duck hawks, too. Around their nests they feed heavily on blue jays and other small woodland birds and on feral pigeons and mourning doves. In coastal environments, especially in late summer, fall, and winter, they feed on a variety of shorebirds and on some ducks, as well as on pigeons.

The peregrine is a well known generalist and opportunist, and it appears that there is sufficient behavioral and physiological plasticity built into the phenotype, so that adaptive adjustments to specific environmental conditions can be made regardless of the precise allelic composition of the genotype. This plasticity should make it much easier to establish a founding population from exogenous sources than would be the case were the genomes highly selected to fit specific environmental configurations.

When possible, it probably is wisest to release only birds from indigenous stock, assuming that a large enough sample exists to avoid problems of inbreeding, as we are doing in our Rocky Mountain peregrine program (Burnham et al. 1978). This rule can be carried to gross and unnatural extremes, however, when, for example, people insist that only peregrines from New Mexico should be released in New Mexico, or only Alberta birds in Alberta, Swedish birds in Sweden, and, of course, only

Falco peregrinus germanicus in Germany—as though these political boundaries have some deep biological signficance to peregrines. It could be that only falcons from western montane habitat should be released in the mountains of Colorado, or only coastal—dwelling peregrines on the sea-cliffs of California; but even this "ecotypic principle" appears to have exceptions, judging from our experience in releasing a variety of ecotypes in the eastern states.

Ornithologists and conservationists must learn to deal with the world the way it really is. It is human-dominated, and it will only become more so. There is no such thing as a "natural ecosystem," if by that term one means an ecosystem unaffected by man's activities and by the introduction of exotic plants and animals. Many forms of life face extinction in the next two decades. In our efforts to maintain the maximum possible diversity of living things, we should not exclude the intelligently considered and careful introduction of exotics into new areas, as Charles Elton (1958) recognized some time ago in his balanced analysis of the invasions of animals and plants.

The results which we have detailed in this report show that the methods that have been developed for the restoration of the peregrine falcon work. On the basis of these results, the Eastern Peregrine Falcon Recovery Team has drawn up operational plans for the five years 1981–85. In this period we expect to be able to establish between 20 to 30 breeding pairs of peregrines on man-made structures (towers and buildings) in the mid-Atlantic Coast and Chesapeake Bay region (North Carolina, Virginia, Maryland, Delaware, New Jersey, Pennsylvania, and New York). We further believe that a population of this size and distribution can become self-maintaining through time within the next five years.

In the same period we also expect to make a significant start toward establishing a breeding population on natural cliff eyries in the Adirondack, Green, and White mountains of New York, Vermont, New Hampshire, and Maine, where predation by great horned owls appears to be markedly less severe than in other natural areas where we have tried to release falcons. This region formerly held a rather high density of breeding peregrines, as more than 100 historical eyries are known in these four states. This fact suggests that the environment in these mountains was, as least formerly, better than most other eastern habitats for nesting peregrines, and the presence of so many historical eyries, as well as other potential suitable cliffs, all rather closely spaced, offers advantages for the establishment and spread of a founding population derived from released falcons. The fact that at least 3 pairs of peregrines have been found breeding at historical eyries in Quebec, Maine and New Hampshire not far away is a further indication that this general region is again suitable for falcons; and the presence of such pairs—there may be others—is another reason for concentrating releases in this region in the next five years.

The significance of the peregrine recovery program goes far beyond the restoration of a single species in North America, for the methods that have been perfected for the peregrine can be applied to many other species of threatened raptors, as well as to other birds, and even to animals such as the cheetah (Brand 1980). The pioneering success with bald eagles in New York State is summarized elsewhere in this issue; also the work on Harris' hawks (*Parabuteo unicinctus*) and prairie falcons in California. The white-tailed sea eagle (*Haliaeetus albicilla*) has been reintroduced to the Isle of Rhum in Scotland and should soon be breeding there (Love and Ball 1979). In Europe, the lammergeier (*Gypaetus barbatus*) is being bred to restore a breeding population in the Alps (*The Eyas* 4(2):5,1980). Now, the gravely endangered California condor (*Gymnogyps californianus*) is another species that will no doubt soon be brought under captive management for propagation and eventual return to nature. Adoption of this still controversial plan would have been inconceivable before the successful demonstration with the peregrine falcon.

As the "Global 2000 Report" predicts with awesome credibility, many more species will be added to the endangered and extinct lists in the next two decades, as the remaining natural areas of the world dwindle down to small islands of preserved habitat in otherwise human-dominated landscapes. In his perceptive summary of the Third International Conference on Breeding Endangered Species in Captivity, William Conway (1980) pointed out that, while we face a diminishing future for wildlands and for wildlife, some semblance of nature can still be saved by developing a new and imaginative inter-relationship between captive propagation programs and wildlife reserves. For wild species that can be bred and preserved in captivity it will be possible, in some cases, to repopulate impoverished landscapes creatively with the fragments of nature that have been so preserved. Our work with the peregrine falcon provides an optimistic example of this sort of conservation engineering.

ACKNOWLEDGMENTS

This work has been financially supported by many private individuals and organizations; a complete listing can be found in The Peregrine Fund Newsletter vols. 1–8 (1973–1980). Major funding came from the following: National Science Foundation, World Wildlife Fund, National Audubon Society, Edward John Noble Foundation, Arcadia Foundation, Richard King Mellon Foundation, Atlantic-

Richfield Foundation, North American Peregrine Foundation, U.S. Fish and Wildlife Service, U.S. Army Armament Research and Development Command, U.S. Forest Service, Maryland Department of Natural Resources, New Jersey Department of Environmental Protection, and New York Department of Environmental Conservation.

We have also had much cooperation and help from the Virginia Commission of Game and Inland Fisheries, the Pennsylvania Game Commission, the New Hampshire Fish and Game Department, Vermont Agency of Environmental Conservation, and the Massachusetts Department of Fisheries, Wildlife, and Recreational Vehicles.

Individuals who have been locally helpful with the fieldwork include: Mitchell Byrd, College of William and Mary; Tom Nichols, Virginia Beach; Rod Hennessey and Barry Truitt, Virginia Coast Reserve, Nature Conservancy; Gary Taylor, Maryland Wildlife Administration; F. Prescott Ward and William Russell, Aberdeen Proving Ground; Paul D. (Pete) McLain, New Jersey Division of Fish, Game and Wildlife; Gaylord Inman, Brigantine National Wildlife Refuge; Jim and Betsy Jones at Barnegat Bay; Harrold Frommelt and Tom Geis, AT and T Long Lines Station, High Seas; John Lanier, Jim McGowan and Barbara Hill, White Mountains National Forest; Eugene McCaffrey and Barbara Loucks, New York Department of Environmental Conservation; Brad Snyder and Carl Beard, Mohonk Reserve; Harrison B. Tordoff, Bell Museum of Natural History, University of Minnesota; Paul Nickerson and Mark Fuller, Fish and Wildlife Service; and members of the Eastern Peregrine Recovery Team: Rene Bollengier, Jr., former leader, Eugene McCaffrey, leader, Donald Hagar, James Baird, Bernard Halla (former member), Paul D. McLain, and Malcolm Edwards.

We also thank the following past and present co-workers with The Peregrine Fund, Incorporated: James Weaver, Willard Heck, Jr., Phyllis Dague, Stanley Temple, Steven Sherrod, Tom Maechtle, Mark MacLeod and Victor Hardaswick.

LITERATURE CITED

Barclay, J. H. 1980. Release of captive-produced peregrine falcons in the eastern United States, 1975-79. M.S. Thesis, Michigan Technological University, Houghton. 118 pp.

Bollengier, R. M., Jr., J. Baird, L. P. Brown, T. J. Cade, M. G. Edwards, D. C. Hagar, B. Halla, and E. McCaffrey. 1979. Eastern peregrine falcon recovery plan. U.S. Fish and Wildlife Service. Mimeographed. 147 pp.

Bonney, R. E., Jr. 1979. Wintering Peregrine Falcon populations in the eastern United States, 1940-1975: A Christmas Bird Count analysis. Am. Birds 33:695-697.

Brand, D. J. 1980. Captive Propagation at National Zoological Gardens in South Africa, Pretoria. International Zoo Yearbook 20:107-112.

Burnham, W. A. 1978. The Peregrine Fund's western program for the propagation and reintroduction of Peregrine Falcons. Unpublished report. The Peregrine Fund, Inc., Fort Collins, Colorado.

———, F. P. Ward, W. G. Mattox, D. M. Clement, and J. T. Harris. 1974. Falcon research in Greenland, 1973. Arctic 27:71–74.

Cade, T. J. 1974. Current status of the peregrine in North America. Pp. 3–12 *in* Proceedings of the Conference on Raptor Conservation Techniques, Fort Collins, Colorado, 22–24 March 1973 (Part 6). Raptor Research Report No. 3.

———. 1980. The husbandry of falcons for return to the wild. International Zoo Yearbook 20:23–35.

———, R. W. Fyfe. 1978. What makes Peregrine Falcons breed in captivity Pp. 251–262. *in* Endangered Birds: Management techniques for preserving threatened species (S. A. Temple, Ed.). Univ. of Wisconsin Press, Madison.

———, J. D. Weaver, J. B. Platt, and W. A. Burnham. 1977. The propagation of large falcons in captivity. Raptor Research 11:28–48.

———, S. A. Temple. 1977. The Cornell University falcon programme. Pp. 353–368. *in* World Conference on Birds of Prey, Report of Proceedings, Vienna 1975. (R. D. Chancellor, Ed.). Int. Council for Bird Preservation.

Clement, R. D., Ed. 1974. Peregrine Falcon recovery, proceedings of a conference on Peregrine Falcon recovery, 13–15 February 1974. Audubon Conservation Report No. 4. National Audubon Society, New York.

Corbin, K. W. 1977. Genetic diversity in avian populations. Pp. 291–302. *in* Endangered Birds: Management techniques for preserving threatened species (S. A. Temple, Ed.). Univ. of Wisconsin Press, Madison.

Conway, W. G. 1980. Where do we go from here? International Zoo Yearbook 20:184–189.

Drury, W. H. 1974. Rare species. Biological Conservation 6(3):162–169.

Elton, C. 1958. The ecology of invasions by animals and plants. London: Methuen and Co., LTD. 181 pp.

Enderson, J. H. 1965. A breeding and migration survey of the Peregrine Falcon. Wilson Bull. 77:327–339.

———. 1969a. Peregrine and Prairie Falcon life tables bsed on band-recovery data. Pp. 505-509. *in* Peregrine Falcon populations: their biology and decline (J. J. Hickey, Ed.). Univ. of Wisconsin Press, Madison.

———. 1969c. Coastal migration data as population indices for the Peregrine Falcon. Pp. 275–278. *in* Peregrine Falcon populations: their biology and decline (J. J. Hickey, Ed.). Univ. of Wisconsin Press, Madison.

Fyfe, R. W., H. Armbruster, U. Banasch, and L. J. Beaver. 1978. Fostering and cross-fostering birds of prey. Pp. 183–193. *in* Endangered birds: management techniques for preserving threatened species (S. A. Temple, Ed.). Univ. of Wisconsin Press, Madison.

Ganier, A. F. 1931. Nesting of the Duck Hawk in Tennessee. Wilson Bulletin. 43:3–8.

Grier, J. W. 1976. Predicting the success of raptor reintroductions through deterministic and stochastic models. Paper presented at the Raptor Research Foundation 1976 Annual meeting, 29 Oct.–1 Nov. 1976. Ithaca, New York.

Groskin, H. 1952. Observations of Duck Hawks nesting on man-made structures. Auk 69:246–253.

Hagar, J. A. 1969. History of the Massachusetts Peregrine Falcon population, 1935–1957. Pp. 123–131. *in* Peregrine Falcon populations: their biology and decline (J. J. Hickey, Ed.). Univ. of Wisconsin Press, Madison.

Henny, C. J. 1972. An analysis of the population dynamics of selected avian species. Bureau of Sport Fisheries and Wildlife Research Report No. 1. Government Printing Office, Washington, D.C.

———, and H. M. Wight. 1972. Red-tailed and Cooper's Hawks: their population ecology and environmental pollution. Pp. 229–250. *in* Population ecology of migratory birds. Symposium Volume. Patuxent Wildlife Research Center, Maryland.

Herbert, R. A., and K. G. S. Herbert. 1965. Behavior of Peregrine Falcons in the New York City region. Auk 82:62–94.

Hickey, J. J. 1942. Eastern population of the Duck Hawk. Auk 59:176–204.

———. 1949. Survival studies of banded birds. Unpublished Ph.D. Thesis. Univ. of Michigan, Ann Arbor.

———, Ed. 1969. Peregrine Falcon populations: their biology and decline. Univ. of Wisconsin Press, Madison.

———, and D. W. Anderson. 1969. The Peregrine Falcon: life history and population literature. Pp. 3–42. *in* Peregrine Falcon populations: their biology and decline (J. J. Hickey, Ed.). Univ. of Wisconsin Press, Madison.

Johnston, D. W. 1974. Decline of DDT residues in migratory songbirds. Science 186:841–842.

Jones, F. M. 1946. Duck Hawks of eastern Virginia. Auk 63:592.

Kochert, M. N. 1976. Reproductive performance, food habits, and population dynamics of raptors. Pp. 1–56. *in* Snake River birds of prey Research Project, Ann. Report 1976. U.S. Dept. of the Interior, Bur. of Land Manage., Boise, Idaho.

———. 1977. Reproductive performance, food habits, and population dynamics of raptors. Pp. 1–39. *in* Snake River birds of prey Research Project, Ann. Report 1977. U.S. Dept of the Interior, Bur. of Land Manage., Boise, Idaho.

Koenig, W. D. and F. A. Pitelka. 1979. Relatedness and inbreeding avoidance: counterploys in the communally nesting acorn woodpecker. Science 206:1103–1105.

Lack, D. 1937. The psychological factor in bird distribution. British Birds 31:130–136.

Love, J. A., N. E. Ball, and I. Newton. 1978. White Tailed Eagles in Britian and Norway. British Birds 71:475–481.

MacArthur, R. H. 1972. Geographic ecology: patterns in the distribution of species. Harper and Row, Inc., New York.

Mayr, E. 1963. Animal species and evolution. Harvard University Press. 797 pp.

Mebs, T. 1960. Probleme der Fortpflanzungsbiologie und Bestandsurhaltung bei deutschen Wanderfalken (*Falco peregrinus*). Vogelwelt 81:47–56.

Michell, E. B. 1900. The art and practice of hawking. D. R. Hillman and Son's, Ltd., Great Britian.

Miller, L. 1930. The territorial concept in the horned owl. Condor 32:290–291.

New Jersey State Geologist Report. 1890. Final Report of the State Geologist: mineralogy, botany, zoology. Vol. 2. J. L. Murphy Co., Trenton.

Newton, I. 1979. Population ecology of raptors. Buteo Books, Vermillion, South Dakota.

Peakall, D. B. 1976. The Peregrine Falcon (*Falco peregrinus*) and pesticides. Can. Field-Naturalist 90:301–307.

Peterson, S. R. 1976. Feeding activity and behavior of Prairie Falcons. Pp. 227–240. *in* Snake River birds of prey Research Project, Ann. Report 1977. U.S. Dept. of Interior, Bur. of Land Manage., Boise, Idaho.

Ralls, K., K. Brugger, and J. Ballou. 1979. Inbreeding and juvenile mortality in small populations of ungulates. Science 206:1101–1103.

Ratcliff, D. A. 1980. The Peregrine Falcon. Buteo Books, Vermillion, South Dakota.

Selander, R. K. 1971. Systematics and speciation in birds. Pp. 57–147. *in* Avian biology, vol. 1, D. A. Farner and J. R. King (Eds.). Academic Press.

Sherrod, S. K. and T. J. Cade. 1978. Release of Peregrine Falcons by hacking. Pp. 121–136. *in* Birds of prey management techniques (T. A. Geer, Ed.). British Falconer's Club.

Shor, W. 1970. Peregrine Falcon population dynamics deduced from band recovery data. Raptor Research News 4:49–59.

———. 1976. Mortality of banded Peregrine Falcons that have been held in captivity. Condor 78:558–560.

Smith, D. G. 1970. Close nesting and agression contacts between Great Horned Owls and Red-tailed Hawks. Auk 87:180–171.

Smiley, A. K., Jr., and D. Smiley, Jr. 1930. An unusual Duck Hawk recovery. Bird-Banding 1:144–145.

Snyder, H. A., and N. F. R. Snyder. 1974. Increased mortality of Cooper's Hawks accustomed to man. Condor 76:215–216.

Terrasse, J. F., and M. Y. Terrasse. 1969. The status of the Peregrine Falcon in France in 1965. Pp. 225–230. *in* Peregrine Falcon populations: their biology and decline (J. J. Hickey, Ed.). Univ. of Wisconsin Press, Madison.

Ward, F. P. 1976. International color-banding of peregrines: 1975 status report. Hawk Chalk 15:35–44.

———, and R. B. Berry. 1972. Autumn migrations of Peregrine Falcons on Assateaque Island, 1970–1971. J. Wildl. Manage. 36:484–492.

White, C. M. 1968. Biosystematics of North American Peregrine Falcons. Unpublished Ph.D. Thesis. Univ. of Utah.

Wiley, J. W. 1975. Relationships of nesting hawks with Great Horned Owls. Auk 92:157–159.

Young, H. F. 1969. Hypotheses on peregrine population dynamics. Pp. 513–519. *in* Peregrine Falcon Populations: their biology and decline (J. J. Hickey, Ed.). Univ. of Wisconsin Press, Madison.

The Bald Eagle in the Northern United States[1]

James W. Grier, Francis J. Gramlich, James Mattsson,
John E. Mathisen, Joel V. Kussman, James B. Elder,
and Nancy F. Green

THE bald eagle (*Haliaeetus leucocephalus*) is a large, long-lived bird of prey restricted in distribution to North America. Adults, with their dark brown bodies, white heads and white tails are well known as the nation's symbol. However, the adult plumage is not acquired until age 4 at the earliest. Bald eagles go through a series of plumages prior to attaining adult coloration, and in some plumages the young bear a superficial resemblance to the golden eagle (*Aquila chrysaetos*).

Sexual maturity is reached at 4 to 6 years of age, but the birds may be considerably older before they breed for the first time. Known-age bald eagles in the wild have bred at 4 years (individuals originally released as nestlings into New York) and as late as 7 years (a color-marked bird in Saskatchewan). The average life span is not known, but 30 years is a reasonable estimate of potential longevity under natural conditions. Mortality is thought to be relatively high in the immature age classes but much lower for birds that manage to survive their first year or two. Many birds probably do not reach sexual maturity, and few are likely to live until age 30.

[1] An excerpt from a preliminary draft of the Northern States Bald Eagle Recovery Plan. Management recommendations contained in this article do not necessarily reflect the opinions of other bald eagle recovery teams and, additionally, may be modified in the recovery plan review process.

Nesting bald eagles are associated almost exclusively with lakes, rivers, or sea coasts. Eagles commonly are said to mate for life, but there are few data on this point. Bald eagle nests are large flat-topped masses of sticks with a lining of finer vegetation such as rushes, mosses, or grasses. The nests are primarily in trees, and to a lesser extent on cliffs or (rarely) on the ground. Clutch sizes range from 1 to 3 eggs. Successful pairs usually raise 1 or 2, occasionally 3 young per nesting attempt.

Adults tend to use the same breeding area, and often the same nest, each year. Nesting phenology depends largely on latitude; egg-laying ranges from November in Florida to May in Alaska and northern Canada. The time between egg-laying and fledging is approximately four months. The entire breeding cycle, from initial activity at a nest through the period of fledgling dependency, is about six months.

The breeding range receded during the 19th and 20th centuries. Now, approximately 90% of the 1400 or so pairs remaining in the conterminous 48 states occur in just ten states: Florida, Virginia, Maryland, Maine, Michigan, Wisconsin, Minnesota, Washington, Oregon, and California. In contrast, large numbers of bald eagles continue to nest in parts of Alaska and Canada.

A few adult and immature bald eagles in the temperate latitudes and far north remain in association with nesting areas throughout the year. However, most bald eagles in the interior Canadian provinces and northern United States move south in late summer or early fall, probably in response to changes in prey availability and weather conditions. As a result, thousands of bald eagles are present in the contiguous 48 states from November through March, which we refer to as the wintering period. Approximately 13,000 were counted during the 1981 nationwide, midwinter survey sponsored by the National Wildlife Federation. Because nesting phenology varies with latitude, the wintering period overlaps with the initial weeks of the nesting season in some areas.

Wintering bald eagles occur throughout the country but are most abundant in the west and midwest. They use terrestrial as well as aquatic habitats and feed on whatever organisms they can catch easily or scavenge, including fish, waterfowl, small mammals, and big game and livestock carrion. The largest concentrations of wintering eagles are located where one or more of these food resources are abundant and relatively easy for the eagles to obtain. The tendency for bald eagles to congregate at certain locations during the wintering period is well known, and for years it had been assumed that most of the birds were at concentration areas. However, recent analyses of data collected during the National Wildlife Federation's nationwide, midwinter surveys indicate that perhaps only about 50% of the bald eagles present in the Region are in concentration areas; others are present in hundreds of locations that are used regularly by 1 to 20 birds. Collectively the smaller groupings and individuals probably are equal in importance to the larger concentration areas.

At night wintering eagles often congregate at communal roost trees, in some cases traveling 20 km or more from feeding areas to a roost site. The same roosts are used for several years. Many are in locations that are protected from the wind by vegetation or terrain, providing a more favorable thermal environment. The use of these protected sites helps minimize the energy stress encountered by wintering birds. It also has been suggested that communal roosting facilitates food-finding.

Although it is clear that bald eagle numbers have declined, the rates, progress, and dynamics of the decline are not well understood. We have relatively good information on the numbers and success of nesting birds but little data on the number of non-breeding adults and subadults, what age they begin breeding in the wild, the length (in years) of the reproductive period for adults, and the turnover and replacement of mates. We have limited information on the population dynamic aspects of wintering eagles and, aside from a small sample of color-marked birds and the first-year recovery rates of banded nestlings, virtually no information on survival rates.

Hypothetical modeling of bald eagle populations, using a variety of models and reproduction-survival schedules, shows that both reproduction and survival are important. Of the two, changes in survival have more impact on the population than similar changes in reproduction. Depending on survival, it is possible for populations with lower reproduction to do better than others with higher reproduction. Reproduction of bald eagles appears to have increased following a period of reduced productivity from the late 1940s to the early 1970s. We do not know for sure what that means without concurrent survival information. The implications are that variation in reproduction may not be quite as important as we formerly thought (although it still is important and cannot be ignored) and that we should pay more attention to survival and the habitat upon which survival depends.

We need to pursue better information and understanding of eagle population dynamics. In the meantime we have to rely on information about numbers of nesting birds During the mid-20th century a new problem, environmental contamination, entered the picture and caused further significant declines in the remaining nesting populations.

Loss of habitat is perhaps the most serious negative factor, certainly the most difficult to halt and reverse. The destruction of wild areas through land development and increased human activity is affecting adversely the suitability of both breeding and wintering areas. The cumulative aspect of habitat loss is the core of the problem. While actions or developments that detrimentally affect individual areas do not appear to "jeopardize the species" as a whole, the cumulative effect of many small, seemingly inconsequential actions on eagles may be significant.

Disturbance, although difficult to assess and evaluate, has been suggested as a cause of reproductive failure in some breeding areas and a factor that adversely affects the suitability of wintering areas. Eagles vary in their response to human activity, some individuals being tolerant, while others are easily disturbed.

Shooting continues to be one of the more serious mortality factors as evidenced by the frequency of gunshot victims handled by raptor rehabilitation clinics and necropsy data published by the U.S. Fish and Wildlife Service. Rehabilitation and autopsy cases show that death of eagles accidentally caught in open-bait or shoreline traps set for predators or fur-bearers is another serious problem in parts of the northern United States.

Eagle population losses from habitat destruction, shooting, or trapping, while severe, usually are identifiable and could be reversed under sound management. Far more insidious are losses resulting from direct or indirect effects of environmental pollutants or contaminants.

Direct toxic effects of organochlorine insecticides have had severe adverse impacts on bald eagle populations. Dieldrin has been implicated most often in acute poisonings, that is, those resulting in deaths of individual birds. However, it is DDE, a metabolite or breakdown product of DDT, that has caused gravest contaminant problems for eagle welfare. Heavy DDT applications were implicated in massive acute kills of birds and other non-target fish and wildlife. By the late 1960s pesticide researchers had discovered and proven experimentally that chronic exposure even to low levels of DDE inhibits reproduction in many bird species. The inhibition results primarily from thinning of eggshells causing failure to hatch. Through physiological mechanisms not fully understood, DDE interferes with calcium metabolism.

Eggshell thinning occurs most commonly in flesh-eating birds, especially those that feed heavily on birds or fish at the ends of long food chains. Eagles living even part of the year in areas with high background levels of DDT absorb amounts sub-lethal to adults but sufficient to cause eggshell thinning and loss of annual production.

Nesting pairs under observation in Maine, New Jersey, New York, and other northern states failed year after year to produce young. Analyses of unhatched eggs disclosed high DDE residues and resultant shell thinning. With curtailment of use of DDT and other organochlorine insecticides in the early 1970s, the problem gradually is being reduced.

Of the heavy metals found in eagle foods, only mercury and lead have been implicated in eagle deaths. Pollution control efforts have reduced the threat from mercury contamination, but we still do not know about lead. In recent years wintering eagles have concentrated near public hunting areas in the fall where they feed on crippled ducks and geese. Secondary poisoning from eating lead-poisoned prey is a growing problem.

Pollutant or contaminant effects may be indirect, as when habitat components are damaged or destroyed, or direct, as when the eagles suffer chemical injury. Indirect effects attributable to pesticides, heavy metals, or the better known industrial pollutant-contaminants generally have not been separable from other, more gross habitat disturbances. When eagles have been driven from historical ranges by human encroachment, it is moot whether there was a concomitant chemical-caused lowering of the fish food base or loss of nesting or roosting trees.

However, one indirect chemical effect that soon will be separable, if not already, is the phenomenon known as acid rain. Hundreds of Northern Hemisphere lakes, notably in Scandanavia and in New York's Adirondack Mountains, have become so acidic that they no longer support viable fish populations. Lakes throughout New England, and the northern regions of Minnesota, Wisconsin and Michigan are considered most vulnerable to acidification. Oxides of sulfur and nitrogen are primary ingredients of acid rain. Stationary and transportation-related burning of fossil fuels are primary sources. Many lake areas already damaged or susceptible to acid rain damage are in wilderness or semi-wilderness forest areas already supporting eagles or with potential for re-introduction. Early indications are that until the acid rain problem is countered successfully, the future is bleak for eagles and other aquatic-based biota.

PRESENT DISTRIBUTION AND ABUNDANCE

Nesting bald eagles have been extirpated in Indiana, Kansas, New Hampshire, Nebraska, and Utah. Evidence of possible natural re-establishment by bald eagles in 1980 exists for Connecticut and Massachusetts. There is no evidence that bald eagles ever nested in Rhode Island or Vermont.

The county distribution of occupied breeding areas between 1960 and 1980 for the north central, central, and northeastern United States is shown in Figure 1. Distribution varied somewhat during this period. Nesting population densities within certain counties declined appreci-

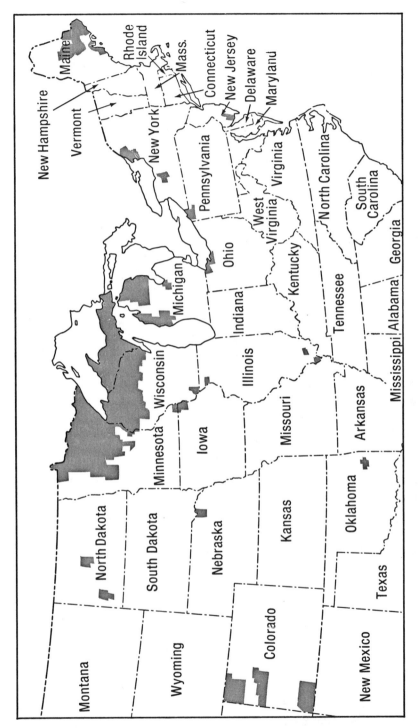

Figure 1. Counties which held occupied bald eagle nesting areas during 1960–1980.

Table 1. Bald eagle productivity[a] in the Northern States Region during 1981[b]

	Number of known breeding areas			Young/nest	
State	Known occupied	Known successful	Young produced	Occupied	Successful
Colorado	5	3	5	1.00	1.67
Connecticut	0	0	0	0	0
Illinois	2	1	2	1.00	2.0
Indiana	0	0	0	0	0
Iowa	1	0	0	0	0
Kansas	0	0	0	0	0
Maine	64	34	49	0.77	1.44
Massachusetts	0	0	0	0	0
Michigan	102	63	105	1.03	1.67
Minnesota	190	132	242	1.27	1.83
Missouri	2	0	(2 reintro.)	0	0
Nebraska	0	0	0	0	0
New Hampshire	0	0	0	0	0
New Jersey	1	0	0	0	0
New York	2	0	(23 reintro.)	0	0
North Dakota	0	0	0	0	0
Ohio	6	3	6 (3 reintro.)	1.00	2.0
Oklahoma	1	0	0	0	0
Pennsylvania	4	3	4 (1 reintro.)	1.0	1.33
Rhode Island	0	0	0	0	0
South Dakota	0	0	0	0	0
Vermont	0	0	0	0	0
Utah	0	0	0	0	0
Wisconsin	202	137	227	1.12	1.66
Total	582	376	640	1.10	1.70

[a]Includes a few artificially reintroduced and translocated young.
[b]Most recent year for which data are collected on regionwide basis.
[c]Large nestlings or fledglings at time of latest survey.

ably but recently have increased in others. In Michigan, for example, during the late 1950s bald eagles nested in 44 counties. A steep decline in reproduction was then already well underway; at least 5–7 counties had been vacated since about 1945. From about 1960 to 1973 the bald eagle disappeared as a breeding bird from 18 (41%) of the 44 counties (including Isle Royale and the Beaver and Manitou Islands). Beginning in 1974 bald eagles re-established themselves in six of these counties, and a nesting attempt occurred in 1981 in a seventh county. This leaves 11 (25%) of the 44 counties where eagles nested two decades ago currently with no nesting pairs.

Productivity data for 1981 are presented in Table 1. Of 582 known

Table 2. Bald eagles counted in the Northern States region during the National Wildlife Federation midwinter bald eagle surveys, 1979–1981

State	Number counted		
	1979	1980	1981
Colorado	316	595	536
Connecticut	20	11	26
Illinois[a]	149	599	405
Indiana	3	5	6
Iowa[a]	41	128	202
Kansas	165	324	308
Maine	109	107	107
Massachusetts	8	25	33
Michigan	30	37	44
Minnesota[a]	3	16	8
Missouri[a]	178	948	955
Nebraska	204	442	440
New Hampshire	0	3	8
New Jersey	6	13	9
New York	41	36	35
North Dakota	30	40	54
Ohio	6	7	10
Oklahoma	581	569	542
Pennsylvania	5	11	26
Rhode Island	1	0	1
South Dakota	62	407	372
Utah	627	661	743
Vermont	0	0	1
Wisconsin[a]	53	70	88
Mississippi River[b]	1,350	945	1,098
Total	3,988	5,999	6,057
Nationwide Total[c]	9,815	13,046	13,527

[a]Does not include eagles counted along the Mississippi River.
[b]From approximately Minneapolis, Minnesota, to Hickman, Kentucky/Dorena, Missouri.
[c]Results include pairs occupying breeding areas in southern states.

occupied breeding areas, 494 (85%) occurred in Minnesota, Wisconsin, Michigan, and Maine. An examination of historical records reveals that the proportion of breeding areas outside these 4 states has decreased markedly. In 1981 only 66 (10%) of 640 young were produced in states other than the 4 mentioned.

The best wide-scale information for the wintering period is from the nationwide midwinter (January) survey coordinated by the National Wildlife Federation. Results of the 1979–81 surveys are presented in Table 2. These data are not directly comparable between states or years because survey coverage is variable. Nevertheless, the data probably do reflect the use of the various states by bald eagles during January.

Vermont and Rhode Island have no known records of former use by wintering eagles, and there is little current use. Wintering bald eagles occur in relatively low numbers in New Hampshire, Connecticut, Massachusetts, Pennsylvania, New York, New Jersey, Ohio, Indiana, Michigan, Wisconsin, Minnesota, and North Dakota. The mid-winter surveys revealed no major concentration areas in any of these states; they accounted for only about 4% of birds counted in the northern United States during January 1980. Although these states appear relatively unimportant in terms of the total number of birds supported, some sites may be important wintering areas for birds from breeding areas within the Region. Maine and Iowa support noticeably larger wintering populations.

Colorado, Illinois, Kansas, Missouri, Nebraska, Oklahoma, and Utah each support several hundred eagles every winter. Collectively, these seven states plus the Mississippi River accounted for over 90%% of the eagles recorded in the midwinter surveys in the region and nearly 50% of the nationwide total.

PRESENT MANAGEMENT

Aside from legislation, e.g., the Bald Eagle Act of 1940, and related occasional enforcement, management efforts and research interests in bald eagles were extremely limited prior to the early 1960s. The National Audubon Society's Continental Bald Eagle Project, initiated in 1960, was the first organized attempt to assess the breeding population and monitor reproductive success of the species across the United States.

As habitat loss and declines in reproduction became known, interest in eagles increased dramatically among federal and state agencies, universities and private organizations. The first agency to develop a specific habitat management program for protection of bald eagle nests was the U.S. Forest Service. In 1963 buffer-zone constraints were established at all known nest sites on National Forest lands in the Great Lakes Region. Since then the biology and habitat requirements of the species have been researched and management strategies have been implemented in some breeding and a few wintering areas.

The U.S. Fish and Wildlife Service and state agencies also began showing more interest in the species during the same period, often recording eagles during waterfowl surveys or conducting specific surveys for the eagles. The National Wildlife Federation, through a corporate grant, established a Raptor Information Center with special emphasis on the bald eagle during the 1976 bicentennial.

Population monitoring efforts in both breeding and wintering areas, which in the early years were confined largely to the mid-west and east, have increased, and now many large portions of the nation are covered by surveys to locate breeding and wintering areas and to monitor nesting success. In the east and mid-west these surveys are handled cooperatively by the U.S. Forest Service, U.S. Fish and Wildlife Service, state wildlife agencies and private volunteers. In the west other agencies, such as the Bureau of Land Management and Bureau of Reclamation have joined in the overall effort.

Organizations such as the National Wildlife Federation, The Nature Conservancy, and Eagle Valley Environmentalists have been effective in acquiring and protecting some essential habitats, especially wintering areas.

Efforts have been underway through cooperative arrangements between state and federal agencies to bolster or re-establish breeding populations by moving captive-bred young or young from relatively secure populations in the Great Lake states and Alaska to suitable but empty habitat in New York and Ohio.

Contaminant monitoring, captive breeding, law enforcement and population monitoring have been major activities of the U.S. Fish and Wildlife Service. When the bald eagle was declared an endangered or threatened species in 1978 under the Endangered Species Act, U.S. Fish and Wildlife Service responsibility was further extended to include administration of the Act and the attendant actions necessary for the recovery of the species. Five regional bald eagle recovery teams were established, as shown in Figure 2. The Bald Eagle Act and the amended (1972) Migratory Bird Treaty Act provide for national protection of bald eagles. Most states also have laws providing for their protection.

EAGLE HABITAT

Essential habitats are locations that biologists consider necessary for continued survival and recovery of the species. The Secretary of Interior officially can designate essential habitat as critical habitat. The distinction between "essential" and "critical" is a legal one; the same biological criteria are used to define each, but economic values are a major consideration in critical habitat designations. Only official designation of critical habitat by the Secretary carries legal force.

Section 7 of the Endangered Species Act directs federal agencies to

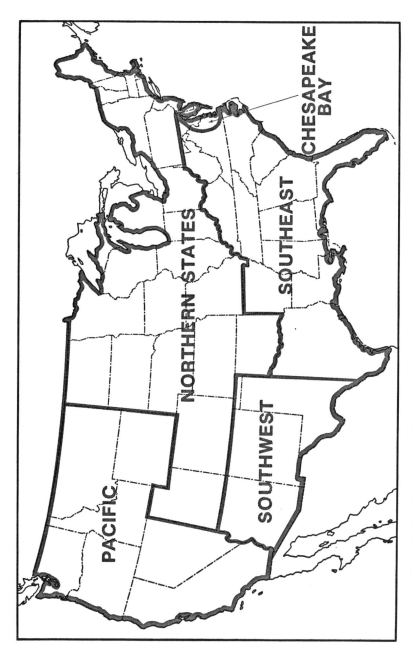

Figure 2. The five regions of the United States that have bald eagle recovery teams assigned to them.

ensure that any action they authorize, fund, or carry out does not result in appreciable adverse modification or destruction of habitat designated as critical by the Secretary. Many federal actions, if properly designed, may take place within the critical habitat of bald eagles (should any be designated in the future) without causing adverse modification.

Title 50, Part 424 of the Code of Federal Regulations defines critical habitat as the specific areas within the geographic area occupied by a species which contain physical and biological features essential to the survival and recovery of the species, and which may require special management consideration or protection. Critical habitat does not normally include the entire geographic range occupied by the species, but may include areas not currently occupied if essential to the recovery effort.

The species requirements considered in the designation of critical habitat are used to identify essential habitat. These requirements include, but are not limited to: space for individual and population growth and normal behavior; food, water, air, light, minerals or other nutritional or physiological requirements; cover or shelter; protection from disturbance and sites for breeding and rearing of offspring. This includes provisions for the future, such as alternate trees for new nests.

For economic reasons, habitats that are believed essential for the survival and recovery of bald eagles might not be designated as critical habitat. The lack of legal designation does not alter the biological importance of essential habitats, and we recommend that these areas receive appropriate management. Preliminary recommended guidelines for identifying essential habitat and managing bald eagles in the region shown in Figure 1 are described below. We stress that these recommendations are subject to change and, furthermore, do not necessarily apply to regions under the purview of other bald eagle recovery teams.

Breeding Areas

We recommend that all potential and occupied breeding areas, including alternate and infrequently occupied nests, suitable historical and other potential locations be considered essential habitat. General guidelines for delineations at each area are as follows:

An initial value, unless or until more specific information is available, for the amount of terrestrial and aquatic habitat considered essential is 640 acres for each breeding area.

The configuration of essential habitat at each site may vary, but should correspond to legal land lines or survey descriptions to facilitate listing and identification in public documents. We recognize the hazard of contributing to the public knowledge of eagle nest locations but believe the location within a 640 acre (or larger) block should not be specific enough to cause significant jeopardy. The actual nest structure(s) does (do) not need to be at the center of the area nor must the area be in any particular configuration (e.g., square).

In areas of high nesting density, a larger single block of essential habitat may be more appropriate than several smaller ones. The essential habitat configuration should be contiguous unless feeding areas or other essential habitat components are relatively far removed from the nesting area. Essential habitat may include private as well as public land. Essential habitat designations associated with breeding areas that become abandoned should remain in effect, provided the sites remain suitable for reoccupation. If the breeding areas are rendered permanently unsuitable the essential habitat designation may be removed.

Essential habitat at unoccupied and potential breeding areas are difficult to evaluate. The job is most easily done by considering areas that are similar to known occupied sites and by consulting with knowledgeable persons.

Areas Used by Non-breeding Eagles

We recommend that areas consistently used by adult or immature non-breeding bald eagles during the breeding or post-breeding period be designated essential habitat. Essential habitat in these areas is not necessarily associated with nest sites. Essential habitat includes terrestrial areas, lakes, coastal shorelines, or river segments associated with important food sources, and a zone for perching, feeding, or roosting that extends a minimum of one-half mile from mean high water elevation. The configuration of essential habitat in each area should correspond to legal land lines or survey descriptions to facilitate listing and identification in public documents. Decisions on the importance of an area, depending on the number of birds involved and availability of other habitat, must be made on a regional basis in consultation with the U.S. Fish and Wildlife Service.

Wintering Areas

Survival of individual bald eagles, particularly those in their first year of life, probably depends heavily on conditions they encounter during the wintering period. The physiological condition of adults at the beginning of each breeding season, an important factor influencing reproductive success, also is affected by how well their energy demands are met in wintering areas. Thus, we believe strongly that recovery of nesting populations in the northern United States depends in part on the eagles having suitable locations to use throughout the wintering period each year.

Although wintering areas unquestionably are important we do not know which particular locations, or how many, are essential for the survival and recovery of the nesting populations in the Region. This is because several thousand eagles from nesting areas in Canada are present in the Region during winter and it is impossible to distinguish between them and birds from threatened or endangered U.S. nesting populations. We have discussed this situation on numerous occasions and have sought the opinions of many research and management biologists who deal with wintering bald eagles. We conclude that to insure the survival and recovery of birds from U.S. nesting populations, significant wintering areas must be protected and managed. Also, we recognize the possibility that some birds move into our threatened or endangered populations bolstering them and serving as an important source of natural recruitment. Thus, the fact that some wintering areas are used primarily by eagles from Canada or Alaskan breeding populations should have little bearing on decisions to protect wintering habitat.

Assuming that the presence of birds indicates the presence of required habitat, as listed above, wintering areas in the Northern States Region that meet any of the following criteria should be considered essential habitats:

1. Locations used by adult or immature wintering eagles known (or strongly suspected) to be from nearby breeding areas.

2. Locations (excluding those along the Mississippi River) used annually by five or more eagles for two weeks or longer in Connecticut, Indiana, Iowa, Maine, Massachusetts, Michigan, Minnesota, New Hampshire, New Jersey, New York, North Dakota, Ohio, Pennsylvania, Rhode Island, Vermont, and Wisconsin.

3. Locations used annually by 15 or more eagles for two weeks or longer in Colorado, Illinois, Kansas, Missouri, Montana, Nebraska, Oklahoma, South Dakota, Utah, Wyoming, or along the Mississippi River.

Our rationale for the first criterion is that the presence of suitable winter habitat near breeding areas reduces the energy expenditure and risks entailed in migration, and could result in increased survivorship for resident eagles of all age classes and higher reproductive success for resident adults.

The second and third criteria deal with areas used by migrant eagles. Hundreds of these areas exist, and there is considerable variation in the use they receive. Ultimately, the importance of any location is determined by its contribution to survival and reproductive success, but there is no way to measure this directly. Instead, consideration is given to factors such as the length of time an area is occupied by eagles each year, the amount of use it receives and its potential for supporting more use, the regularity of use over a period of years or during extreme weather when suitable habitat is most limited, and the number and extent of other wintering areas in the vicinity. Ideally, we would like to compare locations on the basis of a standard index which takes these factors into account. Perhaps such an index can be developed, but at present none exists nor have many areas been studied sufficiently to provide the data for such calculations. Consequently, our criteria are based on limited data and are somewhat arbitrary. The adequacy of these criteria will be reassessed as more data become available.

We consider two weeks, which is roughly 8 to 12% of the wintering period for most migrants, as the minimum period a site must be occupied annually to be considered an essential habitat. Most essential habitats probably are used longer than this, while relatively unimportant areas are used for a shorter time.

Levels of use in the criteria also are somewhat arbitrary and differ throughout the Region. The Great Lakes and Northeast support a low percentage (probably less than 15%) of the winter use now occurring in the Region. Within these states a location that regularly supports 5 or more eagles is quite unusual. Preserving such sites is important for three reasons: (1) suitable wintering habitat should be available throughout the Region, (2) the amount of suitable habitat remaining in the Great Lakes and northeast states appears to be quite limited, and (3) they should be maintained for future use by eagles from the nesting populations that hopefully will be restored in the eastern portion of the region. The situation farther west is different because the numbers of wintering areas and eagles are much higher. In our opinion, any location in the western portion of the Region that annually supports 15 or more eagles is essential habitat and should be managed. We strongly encourage management of sites with fewer eagles, although not at the expense of preserving areas that support more. Further, we urge concerned parties in each state to work together in establishing priorities for inventory and management of wintering habitats.

Guidelines for delineating the boundaries of essential wintering habitats are as follows: The configuration of each essential habitat should include roost sites, aquatic and terrestrial feeding areas, prey habitat, and other biological or physical features necessary for continued use of the site. Essential habitat boundaries should correspond to legal land lines or survey descriptions to facilitate identification. The configuration should be contiguous unless feeding areas and night roosts or other essential habitat components are physically separate. Essential wintering habitat may include private as well as public land.

Our recommendations concerning essential winter habitat are for the Northern States Region only. We are concerned that many adult and immature eagles from breeding areas in the Region rely heavily on wintering areas outside the Region.

MANAGEMENT GUIDELINES FOR BREEDING AREAS

The purpose of these guidelines is to provide criteria for protecting bald eagles at their breeding areas from human disturbance and to preserve and enhance important habitat features of these areas. The criteria are based on a synthesis of existing guidelines in present use by the U.S. Forest Service (Eastern Region), U.S. Fish and Wildlife Service, and the views of eagle researchers.

Although eagles often use particular nests for several years, they move to different sites frequently. Turnover of existing nests, from losses to wind, changes by the eagles, and other natural factors may be as much as 20 percent of the sites per year. Eagle "real estate" is much less fixed than for humans. Thus, the conservation and management of nesting habitat is far more important than the identification and preservation of specific nest sites or even breeding areas.

Eagle tolerance of human presence is highly variable, both seasonally and among different individuals or pairs of eagles. Furthermore, eagle tolerance of humans or, conversely, susceptibility to disturbance, may change from experience, that is, learning. Young birds, particularly, may learn to accept human activity. On the other hand, some eagles may be extremely intolerant and are disturbed readily. All nesting eagles are disturbed more easily at some times of the nesting season than at others.

Four levels of sensitivity to disturbance can be identified for nesting areas. These are as follows.

1. *Most critical.* Prior to egg laying bald eagles engage in courtship activities and nest building. During this and the incubation periods they are quite alert to external disturbances and may readily abandon the area. The most critical period for disturbances therefore extends from approximately one month prior to egg laying through the incubation period.

2. *Moderately critical.* This includes approximately one month prior to the above period and about four weeks after hatching. Prior to the nesting season individual pairs of eagles vary considerably in time of return to the nest site or, if permanent residents, the time they begin to come

into physiological condition for breeding and become sensitive to disturbance. Declaring an earlier period of moderate sensitivity to disturbance provides a buffer period for the early birds and allows for lower, early stages of sensitivity for all birds. After hatching the chicks are quite vulnerable to inclement weather and need frequent brooding and feeding. Disturbance can keep adults from nests and, depending on the weather and length of time involved, may cause weakening or death of chicks. The adults are quite protective of the nest site as long as one or more healthy chicks are present. Thus, disturbance at this time is less critical than during the pre-laying and incubation period.

3. *Low critical period.* This period extends from the time chicks are about one month of age until approximately six weeks after fledging. During this time adults still are quite attached to nesting areas but seem less concerned about moderate amounts of human presence. Again, however, there is individual variation among eagles. In addition, the young have become tolerant of weather conditions likely to occur during this time of year and feeding and attention from the adults have become much less frequent. Premature fledging (and subsequent death of the eaglet, if injured) as a result of human disturbance can be a problem during this period. Short of deaths of the birds (such as from shooting, premature fledging, or accidents), dramatic increases of human activity, or continuous disturbance, human disturbance during this part of the reproductive cycle is not likely to seriously affect the birds.

4. *Not critical.* The existence of this period depends on whether adults are permanent residents in their nesting areas. In most regions adults leave the vicinity for at least a few weeks each year. During the time they are gone they obviously cannot be disturbed and one need be concerned only with activities that alter the habitat in ways that would make it unsuitable for future nesting.

The timing of the above periods depends on geographic location. Eagles tend to breed earlier farther south or in coastal locations. Establishment of critical periods in management planning will therefore depend on the general timing of nesting in each area.

Management of nesting areas will depend on the numbers of pairs present, extent of the areas used by nesting eagles, and present land uses. Plans should be prepared for each site if possible, and planning should encompass the entire region when habitat is suitable and many nesting pairs are present. For planning for a large region, particularly if major changes in land use or development are anticipated, the following major items should be addressed:

1. *Distribution of human development.* If possible, blocks of suitable habitat should remain undeveloped, not just small, specific sites where nests currently are located.

2. *Upper limit to development.* Limits on development should be clearly established in advance, and unplanned development should be discouraged or prohibited. Limits set in advance are generally more acceptable to persons desiring further development; the process permits reasonable negotiation and compromise and limits are easier to enforce.

3. *Rate of development.* Development should only be allowed to approach the upper limit slowly, over a period of years. Hopefully, this will permit raising new generations of more tolerant eagles, as described above. Sudden, large-scale development should be prevented if possible.

4. *Seasonal timing of development.* Construction and other development activities should be confined to the least critical periods of the year, as described above. Some activities can be allowed during the "low critical" period, but this will depend on the type, extent, and duration of the action, as well as its proximity to the nest.

5. *Human attitudes toward eagles in the area.* Much human-eagle interaction depends on the predominant attitude of human residents of each area. Residents and visitors in some areas are very favorably disposed towards the birds, often proud and quite protective. They may be careful not to disturb the birds and may help prevent disturbance or destruction by other persons. Such attitudes should be encouraged through education and law enforcement. Illegal shooting of eagles, especially young birds of the year still in the vicinity of nests during the fall hunting season, should be severely penalized.

The above points pertain to general regions where several eagles may be nesting. The following points pertain to specific nesting sites.

The first site-specific consideration involves essential habitat. That habitat and associated management areas should be tailored to the size and configuration of each location. Within each area, however, one may wish to establish zones of restrictions and enhancement around existing nests. Guidelines for these zones, based on those developed by the U.S. Forest Service in the Eastern Region and in several parts of the United States, are described below. If buffer zones are used they should be established around all nest sites within a breeding area regardless of their activity status, since alternate nests often are used as feeding platforms and roosting sites.

Buffer Zones for Nest Trees

Each nest within a breeding area is protected by three zones that become less restrictive to human activity as the distance from the nest increases. Some activities need be restricted only during the nesting season, or critical periods.

Primary Zone

The boundary of this zone should be 330 feet from the nest. All land-use, except actions necessary to protect or improve the nest site, should be prohibited in this zone. Human entry and low-level aircraft operations should be prohibited during the two highest critical periods unless performed in connection with eagle research or management by qualified individuals. Restrictions on human entry during the low critical period should be decided on a case-by-case basis taking into account the type, extent, and duration of the activity. Care must be taken to avoid premature fledging during the last few weeks eaglets are in the nest. Any motorized access, except by boat, into this zone should be prohibited.

Secondary Zone

This zone should extend 660 feet from the nest. Land-use activities that result in significant changes in the landscape, such as clearcutting, land clearing, or major construction, should be prohibited. Actions such as thinning tree stands or maintenance of existing improvements can be permitted, but not during the two highest critical periods. Human entry and low-level aircraft operations should be prohibited during the most critical period unless performed in connection with necessary eagle research and management by qualified individuals. Roads and trails in this zone should be obliterated, or at least closed during the two highest critical periods. Restrictions on activity during the low critical period should be decided on a case by case basis.

Tertiary Zone

This is the least restrictive zone. It should extend one-quarter mile from the nest, but may extend up to one-half mile if topography and vegetation permit a direct line of site from the nest to potential activities at that distance. The configuration of this zone, therefore, may be variable. Most activities are permissible in this zone except during the most critical period. Restrictions on activity during the moderate and low critical periods should be made on a case by case basis. Each breeding area management plan may identify specific hazards that require additional constraints.

Abandoned Nest Trees

When a tree containing an eagle nest has blown down or has been damaged so it can no longer support a nest, all buffer zones should be removed. When a nest structure disappears, but the nest tree remains, the buffer zones should remain in effect through at least the following three breeding seasons. In either case (nest and tree destroyed, or just the nest) if the nest is not rebuilt, the zoning can be removed but the area should still be considered as essential habitat and protected accordingly.

When a nest is classified as a remnant, that is, one that has been unoccupied for five consecutive years, and is not being maintained by eagles the zoning can be removed.

Roosting and Potential Nest Trees

Three or more super-canopy trees (preferably dead or with dead tops) should be identified and preserved within one-quarter mile of each nest as roosting and perching sites. In areas identified as potential nesting habitat, there should be four to six over-mature trees of species favored by bald eagles for every 320 acres within 1/4 mile of a river or lake larger than 40 acres. These trees should be taller than surrounding trees or at the edge of the forest stand, and there should be clear flight paths to them.

Artificial nest structures may be provided where suitable nest sites are unavailable in occupied or potential habitat. Structures may be placed in trees containing dilapidated nests; in trees without existing nests, but which otherwise appear suitable; or in man-made structures such as powerlines or tripods. Nest platforms should be approximately 5 to 6 feet in length and width (25–36 square feet) and be made to last for several years. Roosting structures may be erected using powerpoles with several horizontal perches near the upper end.

Prey Base Management

Fisheries management should strive to maintain a prey base consistent with eagle food habits. In some breeding areas, particularly in the west, mammals form a portion of the diet of bald eagles. Land management in these areas should maintain an adequate prey base in terrestrial habitats. Feeding of eagles may be considered a valid management tool in areas where natural prey are highly contaminated or temporarily unavailable for some reason. This management option rarely will be used. In some regions, commercial and sport fishermen may be providing an important but undocumented food source for eagles by dumping rough fish. Many commercial fishermen are suffering because of reduced catches of game fish and quotas imposed for the purpose of managing fisheries.

Subsidization might benefit eagles, fishermen, and possibly the fisheries.

MANAGEMENT GUIDELINES FOR WINTERING AREAS

Wintering bald eagles have not received the wide-spread, long-term attention given to nesting bald eagles. Information on wintering birds for the longest period of years comes from the region along the Mississippi and associated rivers, particularly in Wisconsin, Iowa, and Illinois. The information consists chiefly of counts of birds and records of wintering locations. A few detailed studies of night roosts, feeding areas, and eagle movement patterns have been conducted at scattered places, particularly in Illinois, Wisconsin, Missouri, Utah, Colorado, South Dakota, and Nebraska. However, these generally have been short-term studies concerned with limited geographic areas, and much of the information has not been synthesized and reported.

In addition to the paucity of information on wintering eagles, the birds use a much wider variety of habitat than when nesting; their behavior is much more variable in their use of that habitat; and they are much less faithful in their use of particular sites. That is, some wintering sites are used only for short periods and the eagles may or may not return to the same site in subsequent years. This apparently depends on factors such as weather, quantity and concentration of food, availability of alternate locations, and human disturbance. At nesting sites adult eagles invest considerable effort and resources in construction of nests, laying and incubation of eggs, and rearing of young. Such is not the case with wintering sites. One would not, therefore, expect bald eagles to evince as much site tenacity at wintering areas.

As a result of both lack of information and variability of habitat use by wintering eagles, less is understood about what wintering birds require, and management is more difficult. The object of management at wintering areas is to maintain or improve their suitability for bald eagles. Because we do not know yet the best ways to accomplish that, we can provide only general guidelines for starting the task. We recommend more research into the needs of wintering eagles and better communication among persons attempting to manage the birds at this time of year.

Management of wintering eagles should focus, at least initially, on areas that are known to be used consistently each year by concentrations of birds. Therefore, the first step is to conduct surveys of numbers and specific locations of birds. In any geographic region, such as a state or an expanse of land under the management of a particular agency, the most important sites will be those where the greatest number of birds are found over the longest periods of time.

The second step, after determining which areas are being used consistently by large numbers of eagles, is to begin thinking about site-specific plans for the protection and management of those areas. The most important considerations will depend on whether a particular site is a night roost or a feeding and daytime use area. For night roosts, the prime considerations should be maintenance of roost trees, prevention of human disturbance, both while the birds are present and over the long term. Disturbance when eagles are arriving or present at a roost may cause them to abandon an area altogether. Long term activities, including some when the birds are not present, may alter the suitability of the habitat for future use. At feeding sites the prime consideration in most cases will be continued availability of food, but prevention of human disturbance is a second important consideration.

The next step involves consideration of known historical and present circumstances and characteristics of a site. If bald eagles are using an area, initial management should be directed toward maintaining present conditions. Changes should be made only after careful deliberation and knowledge of what can be done to improve the area for eagle use. If the birds are using areas away from human disturbance, increased human activity should not be permitted unless it is consistent with the birds welfare. Other areas, however, may be situated in the vicinity of much human activity. Some feeding and daytime use areas in the midwest and east, for example, are located below dams in or at the edge of towns and cities. In such cases, where the eagles clearly are accustomed to human presence and activity or in areas that are used less consistently by the eagles, management may be less restrictive.

In some cases, management plans may incorporate some form of zoning, with the intensity of restriction and management varying with distance from the site, as has been used for nest sites. The zone approach should not distract from the real needs of the birds, however, and, in general, we recommend site-specific plans with size and shape tailored to the particular location. Easements, cooperative agreements with land owners, or acquisition from willing sellers might be necessary for management of most privately owned sites. Management on public lands should be emphasized in locations of mixed ownership.

After considering past and present circumstances and deciding the boundaries of the area to be managed, plans need to address two broad categories of management: restrictions and enhancement. The following points identify some of the items to be considered for night roosts and feeding areas respectively.

MANAGEMENT GUIDELINES FOR NIGHT ROOSTS

Management areas for night roosts should encompass the tree(s) in which eagles actually spend the night, trees used for perching during arrival or departure, and other trees or physical features such as hills, ridges, or cliffs that provide wind protection. Flight corridors regularly used by eagles moving to or from roosts should be included in the plan. In all cases land use decisions must consider the nature of the action, its spatial relationship to the roost, and the current level of disturbance in the area.

Human activity should be restricted during the time of year when bald eagles are present. Exceptions can be made for activities involving eagle research or management by qualified persons and for existing activity being tolerated by roosting eagles. The extent to which bald eagles at roosts tolerate human activity is now known, nor if there much information concerning the distance at which the presence of humans becomes disturbing. Pending further research on these topics, human activity should be discouraged or prohibited within a one-quarter mile radius around a roost when eagles are present. Limited restrictions may be necessary out to one-half mile if there is a direct line of vision from the roost to potential activities.

At locations where eagles already have habituated to a high level of human activity even relatively loud, noticeable disturbances, such as road repair, may be tolerated, particularly if such activities take place only during the day when most eagles are away from roost. However, at every site there probably is a threshold of disturbance which, if exceeded by the cumulative effect of several activities, will cause abandonment of the roost by some or all of the eagles using it. Therefore, human activity should be minimized to the fullest extent possible, or allowed with the stipulation that it must cease if it disrupts use of the roost site.

Occasional activity that does not permanently affect the suitability of the site for roosting can be allowed, provided the activity is brief (e.g. one to five hours), takes place during the time of day when no eagles are present, and that there is sufficient monitoring to insure that the activity does not disrupt use of the site by eagles.

Land use that would result in the destruction of trees in the roost area should be prohibited. Alteration of physical features such as cliffs, or ridges should be prohibited if the alterations would significantly lessen the visual screening or the wind protection these features provide. Construction of highways, roads, railroads, gravel pits, mines, buildings, airports, or other structures should be prohibited. Use of the area by livestock should be controlled or prohibited if it is determined that such use prohibits regeneration of roost trees.

Immediate threats to trees should be minimized, e.g. stabilize banks, control erosion, place protective screening or fencing around trees in areas where damaged by beavers, livestock, or other animals is a problem. Young trees should be planted in locations where natural regeneration is not sufficient to provide roosting, perching, or wind-buffered trees on a sustained basis, or where additional buffering from human disturbance is desired. Also, new roosts could be created by planting trees in suitable wind-protected sites where no trees currently exist. Openings that allow easy access to roost or perch trees in dense stands should be maintained. Access to potential roost or perch trees in dense tree stands should be provided by creating openings or selectively thinning.

MANAGEMENT GUIDELINES FOR FEEDING AREAS

Habitat for prey species should be maintained or improved. Actions that are likely to reduce the abundance or availability of suitable prey to such an extent that fewer eagles would be supported in any area should be prohibited. This is the most important management consideration in feeding areas.

Human access, particularly in areas of concentrated use, should be restricted during months when bald eagles are present. Exceptions could be made for activities involving eagle research or management by qualified persons, current activities being tolerated by eagles, and occasional activities of short duration (e.g. repair of telephone line, checking stockponds). Within large concentration areas there might be small areas which, for a variety of reasons, rarely are used and have low potential for future use by bald eagles. Human activity probably could occur in such places without significantly affecting bald eagles.

Land use that would destroy, or otherwise make unsuitable, trees or other habitat features used as hunting or resting perches should be prohibited. Similarly, vegetation or physical features that screen feeding areas from human activity should be maintained. At locations where suitable perches are in limited supply, sites where young trees can be planted should be identified and protected. Also, it may be necessary to control the use of feeding areas by livestock, if such use is preventing tree regeneration. The use of toxic materials in local rodent or predator control and exposed-bait furbearer traps should be prohibited in eagle concentration areas.

The number of hunting and resting perches can be increased by planting trees. This is especially important in locations where suitable perches are few in number and in locations where tree regeneration is insufficient to maintain perches on a sustained basis. When planting trees keep in mind that trees within 100 feet of water, especially those with limbs very near or hanging over water, are preferred perches in aquatic habitat. In terrestrial habitats preferred perches are trees with commanding views of prey habitat; for example, solitary trees or tree on edges of woodlots or forests.

Existing trees may be made suitable for use as perches by girdling a few limbs or entire trees to create bare branches (this practice is seldom needed for deciduous trees), or by creating a clear flight path to suitable perches by cutting selected branches or trees. Artificial perch structures for bald eagles have met with very limited success. However, artificial perches might be used at locations where prey are available but perch trees are few in number.

SUMMARY

The most important problems believed to have reduced bald eagle populations in the north central, central, and northeastern United States have been: (1) loss of suitable habitat, (2) mortality from shooting, accidental trapping, lead poisoning, and other sources, and (3) reduced reproduction caused by environmental contaminants. Incomplete and inconsistently-reported information plus inadequate communication and coordination among agencies and individuals working with eagles have contributed to problems in our understanding of the species' population dynamics, status, and requirements.

The most important needs for recovery and management of the species in this region of North America are:
1. obtaining better population and habitat data,
2. identifying and managing essential habitat.

These needs apply to both nesting and wintering areas.

Reestablishment of self-sustaining breeding populations in many states is expected to occur by the natural expansion of existing populations, provided suitable habitats are maintained. However, where populations have been extirpated or severely reduced, restoration can best be accomplished by transplanting wild- or captive-produced young to suitable locations.

The ultimate success of efforts to restore breeding populations, whether by natural or artificial means, depends on survivorship. Providing improved habitat conditions, particularly during the winter period, probably is the most important means of maximizing survivorship. Other steps include the development and implementation of programs to reduce deaths from shooting, accidental trapping, electrocution, lead poisoning, or exposure to various environmental contaminants. High priority is given the the rehabilitation if sick or injured eagles, in part because of the public education associated with such programs.

It is imperative that surveys, research, and management planning be carried out by experienced, qualified personnel. This is particularly important for all work involving the eagles themselves, such as when birds are captured or handled or when nests are visited.

California Condor Reproduction, Past and Present

Noel F. R. Snyder

THE decline of the endangered California condor (*Gymnogyps californianus*) is still not well understood, despite several decades of intensive study by a succession of field researchers, including Koford (1953), Miller et al. (1965), Sibley (1966–1969), Borneman (1965–1980), and Wilbur (1978). The gathering of useful data has been hampered by the small numbers of condors, the inaccessibility of much of the species' range, and by problems in following individual birds consistently enough to evaluate factors affecting survival and reproduction. Nests have been difficult to locate, and the population has been impossible to census accurately with existing methodology. In the absence of comprehensive information on the major causes of decline, there has been no reliable way to devise effective means for recovery of the species. A variety of protective measures have been tried, but the condor's status has continued to deteriorate.

Wilbur (1980) offered recent estimates of the size of the wild condor population (Figure 1). Although these estimates are of unknown accuracy, they are reasonably consistent internally, as they are based largely upon repeated counts in certain key regions of the species' range. The numbers of birds seen in traditional use areas, such as the Sespe and Sisquoc condor sanctuaries, have declined continuously and steeply over recent decades, and concentrations of birds have not meanwhile appeared in new areas. Thus, regardless of the numerical accuracy of Wil-

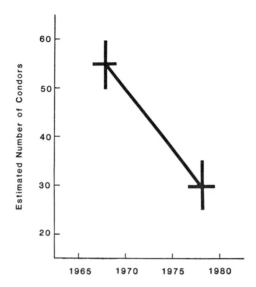

Figure 1. Recent population estimates for the California condor (after Wilbur 1980).

bur's figures, the trend indicated by his estimates is not in question, and there can be no reasonable doubt that the species has been heading rapidly toward extinction. On the basis of counts of the past 2 years and Wilbur's earlier figures, we have been estimating only 20 to 30 individuals left in the wild population (see Ogden and Snyder 1981).

The causes of the condor's decline have long been debated without a consensus emerging. Some observers (e.g. Miller et al. 1965) have emphasized mortality factors, while others (e.g. Wilbur 1978) have argued for problems with natality. Among the potentially important mortality factors are losses resulting from shooting, ingestion of lead shot in carcasses, collisions with overhead wires, fouling in uncovered oil sumps, and secondary poisoning with compound 1080 used in ground squirrel (*Citellus beecheyi*) control programs. Among suggested natality problems are poor nesting success resulting from DDT contamination or from calcium deficiencies in diet, and chronic failure of pairs to attempt breeding because of poor food supplies, disturbance or other factors. Which, if any, of the above problems have been major factors in the decline has not been clear from the available data, and it is possible that the most important causes of decline have yet to be hypothesized. Conceivably there may be no dominant factors in the decline, only a variety of negative forces, each contributing a relatively minor impact.

In an effort to evaluate the importance of some of the potential stress factors, I recently carried out a review of available reproductive data for the species contained in the files of the condor research center. In this paper I present some of the highlights of this review and some inferences that can be drawn about current reproduction of the species from intensive field studies by the condor research center staff and cooperators during the past 2 years.

CONDOR NESTING RECORDS

Before Koford began his condor studies in 1939, data on condor nests were limited largely to records of egg collectors, who almost invariably caused the failure of the nests they discovered. These data tell us nothing of value regarding natural nest success of the species. The data available since 1939 (Table 1) are more useful, but the number of nests found and studied has not been large, and most nests have not been followed through the entire breeding cycle. In preparing Table 1, I excluded all reported "nests" that were not substantiated by observations of eggs, young, or persistent parental attendance. Many of the excluded "nests" were clearly only roost sites. On the other hand, I included a number of nestings for which no actual nest caves were located, but whose existence can be reasonably inferred from the presence of recently fledged young in the nesting areas. Observations of intensively studied family groups suggest that recent fledglings generally retain at least an intermittent association with their nests of origin until the summer following fledging, and can be recognized by characteristics of coloration and behavior, especially by clumsy flight, dependent interactions with adults, and extreme unwariness of humans. The inclusion or exclusion of reported nests has required judgments as to the quality of historical records. Documentation is weakest for the nests noted by asterisks. The conclusions that follow regarding nest success are not significantly affected by inclusion of these nests.

NUMBERS OF NESTS LOCATED

The numbers of nests located in various years cannot be considered an accurate reflection of the actual numbers of nesting attempts in the population, primarily because the amount of field work directed toward nest-finding has varied significantly over the years, as has the methodology used in locating nests. Listed in Table 2 are the numbers of nests found per year over the past 15 years as a function of numbers

Table 1. California condor nesting records, 1939 to present

Nest number	Nesting year	Dates observed	Stages observed	Fledging success	Comments
Pre-DDT years					
S151	1939	3/23/39—3/24/40	Egg to fledgling	Successful	
S231	1939	8/15/39—10/26/39	Nestling to fledgling	Successful	
S23	1939	3/16/39	Egg	Failure	Egg taken by collector
S23	1939	9/30/39—12/13/39	Nestling to fledgling	Failure	Broken wing, found dead
No #	1939	Aug/39—7/13/40	Fledgling	Successful	
S235	1940	5/28/40—10/9/40	Nestling	Unknown	
S41	1940	6/11/40—11/29/40	Nestling to fledgling	Successful	
S161	1940	10/17/40	Fledgling	Successful	
No #	1940	11/14/40	Fledgling	Successful	
S233	1940	3/17/40	Egg	Failure	Egg taken by collector
No #	1940	11/30/40—2/18/41	Fledgling	Successful	
S41	1941	3/16/41—5/16/41	Egg	Failure	Broken egg
S232	1941	3/23/41—6/6/41	Egg to nestling	Unknown	
S23	1943	3/21/43	Egg	Failure	Egg taken by collector
S233	1945	4/3/45—10/20/45	Egg to nestling	Failure	Young starved(?) poisoned(?)
No #	1945	3/23/46—8/15/46	Fledgling	Successful	
S231	1946	3/3/46—4/8/46	Egg	Failure	Broken egg
S152	1946	3/17/46—4/12/46	Egg	Failure	Broken egg
S91	1946	2/12/46—7/23/46	Egg to nestling	Unknown	
No #	1946	6/3/46	Egg	Unknown	Eggshell below cliff
S132	1946	1946	Egg to fledgling	Successful	
S341	1946	8/16/46—9/14/46	Nestling	Unknown	
DDT years					
S302	1950	6/5/50—10/15/50	Nestling	Unknown	Nest in big tree
S231	1951	5/27/51—5/30/51	Nestling	Unknown	
S191	1956?	1956?	Fledgling	Successful	
S234[a]	1956	9/3/56	Nestling	Unknown	
S155	1958	10/15/58—2/24/59	Nestling to fledgling	Successful	
S11	1960	4/25/60	Nestling	Unknown	
S41	1964	6/15/64	Egg	Failure	Broken egg
No #	1965	Early mid 1966	Fledgling	Successful	
No #	1965	4/7/66	Fledgling	Successful	
S13	1966	11/30/66	Fledgling	Successful	
S41	1966?	8/15/66	Egg	Failure?	Broken egg? Eggshell possibly from successful nest of 1965
No #	1966	1/4/67—5/4/67	Fledgling	Successful	
No #	1966	1/4/67—5/12/67	Fledgling	Successful	
S94	1966	2/8/67—2/15/67	Fledgling	Successful	Topatopa
S362	1966	10/27/66	Nestling	Failure	Young found dead under nest
S12	1967	5/24/57—12/16/67	Nestling to fledgling	Successful	
S41	1967	7/17/67	Egg	Failure	Broken egg
S156	1967	1/5/67—10/12/67	Egg to fledgling	Successful	
S162	1967	5/11/67—6/20/67	Egg	Failure	Broken egg
S242	1967	8/10/67	Nestling	Unknown	
S251	1967	7/26/67	Egg	Failure	Broken egg
S15	1968	7/30/68—11/30/68	Nestling to fledgling	Successful	
No #	1968	3/14/69	Fledgling	Successful	
No #	1968	5/15/69	Fledgling	Successful	
No #	1968	5/25/69	Fledgling	Successful	
S231	1979	7/23/69	Egg	Failure	Broken egg
S16	1969	7/9/69—3/13/70	Nestling to fledgling	Successful	
No #	1970	11/19/70—7/3/71?	Fledgling	Successful	
No #	1971	12/15/71—5/31/72	Fledgling	Successful	
No #	1971	9/17/71	Nestling (probable)	Unknown	

Table 1. (continued)

Nest number	Nesting year	Dates observed	Stages observed	Fledging success	Comments
Post-DDT years					
S313	1972	1/4/73—4/2/73	Fledgling	Successful	
S231	1972	9/14/72—5/2/73	Fledgling	Successful	
No #	1972	Early spring 1973	Fledgling	Successful	
No #	1972	April 1973	Fledgling	Successful	
No #	1974	5/13/75	Fledgling	Successful	
S231	1974	11/12/74—3/21/75	Fledgling	Successful	
No #	1975	12/24/75—6/29/76	Fledgling	Successful	
No #	1975	3/24/76—4/21/76	Fledgling	Successful	
S157	1975	10/28/75	Fledgling	Successful	
S135	1976	10/23/76—12/21/76	Nestling to fledgling	Successful	
No #	1976	12/21/76	Fledgling	Successful	
No #	1976	4/19/77	Fledgling	Successful	
No #[a]	1977	10/18/77	Fledgling	Successful	
S135	1977	3/11/77—1/8/78	Egg to fledgling	Successful	
No #	1979	Spring-summer 1980	Fledgling	Successful	
S353	1980	3/4/80—7/22/81	Egg to fledgling	Successful	
S62	1980	6/14/80—6/30/80	Nestling	Failure	Handling loss of nestling
S33[a]	1980	3/12/80—6/5/80	Egg to nestling (?)	Failure	Actual nest not located until after apparent failure
No #[a]	1980	5/15/81—9/2/81	Fledgling	Successful	
S354	1981	6/10/81—present	Nestling to fledgling	Successful	
S303[a]	1981	2/16/81—4/1/81	Egg (?)	Failure	Actual nest not located until after apparent failure
S135[b]	1981	4/22/81—6/23/81	Egg to nestling	Failure	Chick dies close to hatching, cause unknown

[a]Validity of nest record in some question.
[b]Same pair as at S303 in 1981.

of field days spent per year in the condor range by biologists working directly or cooperatively with the condor research program. While biologist-field-days per year is only a rough measure of the amount of effort put into nest-finding, it would be difficult to develop a more rigorous measure. Use of biologist-field-days at least provides a crude indication of the large changes that have occurred over the years in the exposure of competent researchers to potential nest-finding situations. Almost without exception, the condor nests found since 1966 have been located by the biologists whose field days are included in Table 2.

It is clear that the amount of field time per nest found has increased greatly over the past 15 years, a result reinforcing the population decline estimates of Wilbur (1980). While an apparently large number of nests (4) was located in 1980, the number of biologist-field-days increased enormously in this year, as did the number of field days per nest found. Similarly, while three nesting pairs have been located in 1981, the number of biologist-field-days has been running at about the same rate as in 1980. These findings urge restraint in concluding that the numbers of

Table 2. Biologist-field-days and numbers of nests found, 1966 to 1980

Variable[a]	1966–70	1971–75	1976–80	1980 only
Mean field days per year (including surveys)	380	286	625	1,558
Mean field days per year (excluding surveys)	231	215	446	1,303
Mean nests found per year	4.2	2.0	2.0	4.0
Mean field days per nest (including surveys)	90	143	313	390
Mean field days per nest (excluding surveys)	56	108	233	326

[a] Surveys include the annual October Survey and Mini-surveys run by J. Hamber. Field days included in the table were of biologists F. Sibley, J. Borneman, D. Carrier, S. Wilbur, R. Smith, J. Hamber, D. Van Vuren, Y. Miller, D. Connell, B. Walton, F. Baldridge, E. Johnson, N. Snyder, J. Ogden, and their assistants. Surveys included many additional personnel, especially from the U.S. Forest Service and the California Department of Fish and Game. Data from 1981 not included.

nests found in the past two years indicate incipient recovery of the population.

DISTRIBUTION OF NESTS

The searching of known nesting areas for signs of reproductive activity has not been a systematic process. Some nests have been discovered on first visits to areas. Others would not have been found in the absence of exhaustive hunts stretching over periods of several months. In many years the known nesting areas have been visited much too briefly to allow conclusive statements about nesting activity. Thus, the nests that have been found have unquestionably represented an incomplete and inconsistent sampling of the nesting activity actually occurring. The most thorough searching of nesting areas has taken place in the last two years.

The distribution of nesting records by major nesting areas known active since 1939 is detailed in Table 3. Known nesting areas have been delineated as separate drainage systems or other natural geographical units containing one or more known nest sites. To protect the confidentiality of nest sites, actual geographical locations are not presented. While the nests and nesting areas listed in the table very likely do not include all nests and nesting areas active in recent years, there has been an unmistakable contraction in the number of nesting areas known active over the years, a tendency that has been especially marked in the Sespe condor sanctuary region. Of the known areas, nine were clearly active

Table 3. Distribution of known or highly probable condor nests, 1939 to 1981.

Nesting area	Total sites known	Status of nesting area during indicated year[a]																	
		1939–46	1947–65	1966	1967	1968	1969	1970	1971	1972	1973	1974	1975	1976	1977	1978	1979	1980	1981
1	1	+	(1)	+	+	+	0	+	0	+	+	0	0	0	0	0	0	+	+
2	3	+	(1)	+	+	+	+	+	0	+	+	+	+	+	+	+	+	+	+
3	1	0	+	+	+	+	+	0	+	+	+	+	+	0	+	+	0	+	+
4	4	(2)	(3)[b]	(1)	(1)	(1)	(1)	+	+	+	+	+	+	+	+	+	+	+	+
5	5	(3)	(1)[b]	+	(1)	(1)	(1)	(1)	+	+	+	+	+	+	+	+	+	+	+
6	4	+	0	(1)	(1)	+	+	+	+	0	+	+	+	+	+	+	0	+	+
7	2	(3)[c]	+	+	+	(1)	(1)	(1)	(1)	(1)	(1)	(1)	(1)	+	+	+	+	+	+
8	10	(8)	(10)[b]	(2)	(1)	(1)	(1)	(1)	(1)	(2)	(1)	+	+	+	+	0	0	+	+
9	3	(1)	+	+	+	+	+	+	+	+	+	(1)	(1)	+	+	+	0	+	+
10	2	+	(2)[b]	+	(1)	(1)	+	+	+	+	+	+	+	+	+	+	+	+	+
11	6	(2)	(3)[b]	(3)	(2)	+	+	+	+	+	+	(1)	(1)	(2)	(1)	(1)	+	(1)	(1)
12	6	(3)	(2)[b]	(1)	+	(1)	(1)	(1)	(1)	(1)	(1)	(1)	(1)	(1)	(1)	+	+	(1)	(1)
13	2	+	(1)[b]	+	0	+	+	0	0	0	0	0	0	(1)	(1)	(1)	(1)	(1)	(2)[c]
14	2	+	0	0	0	0	+	0	0	(1)	(1)	+	(1)	(1)	+	0	0	(1)	(1)

[a] ( ) = Number of nests known active or highly probably active in period under consideration. Active sites were those with eggs, nestlings, or dependent fledglings in attendance. Since nesting normally takes 2 years, successful sites are active for 2-year periods.
+ = Nesting area checked by condor researchers, but no evidence for nesting obtained. Checking very often far too brief and incomplete to qualify as strong evidence for no nesting.
0 = Nesting area not checked for nesting activity by condor researchers.
[b] Includes nests checked in late 1960s by F. Sibley and judged to have been active during this period.
[c] Two apparent nesting attempts in same year by single pair.

between 1966 and 1970, seven were active between 1971 and 1975, whereas only five were known active between 1976 and 1980. Although the five areas known active in the past two years might appear to represent a small increase in reproductive activity from the immediately preceding years, they very likely represent only better coverage, as discussed above. The overall decline in observed use of known nesting areas has paralleled the estimated population decline, and gives support to the apparent downward population trend.

NEST SUCCESS

Ideally, nest success should be evaluated by using nests found at the very start of the nesting cycle and followed through to failure or fledging. Productivity should be computed on the basis of number of young produced per nesting attempt. Unfortunately, hardly any condor nests have been found at the start of nesting, and few have been found by methods that would have allowed detection of early failures had they occurred. In fact, a large fraction of the "nests" found, especially in recent years, have been recently fledged young for which actual nest sites were never confirmed. About the best one can do in attempting to calculate something meaningful about historical nest success is to limit analyses to: (1) nests found at some time during the egg stage, and (2) nests found later in the breeding cycle whose existence would likely have been detected even if they had failed early. Because most failed nests contained evidence of nesting in the form of eggshells, feathers, pellets, and whitewash, early failures can generally be detected if nest sites are closely inspected.

With the above limitations, and the limitation that nests clearly lost to human disturbance should be excluded from analysis, one is left with nests that permit an approximate evaluation of natural nesting success. However, it must be assumed that even this restricted nest sample will overestimate true nest success to some extent because of the inclusion of some nests found late in the incubation period. The number of nests qualifying for analysis is not large enough to give more than preliminary indications of natural nest success over the years.

To examine the possibility that DDT contamination may have adversely affected nest success, I have grouped the nests of Table 1 into three eras. During pre-DDT years 22 condor nestings were documented, but 13 of these must be excluded from analysis because they were nests that failed because of egg collectors or were nests found at the nestling or fledgling stage that probably would not have been detected if they had failed at the egg stage. Of the remaining nine nests, six were followed to a known conclusion, while three had unknown outcomes. Minimum nest success (assuming all nests of unknown outcome failed) was two nests out of nine successful in fledging young (22%); maximum nest success (assuming all nests of unknown outcome succeeded) was five nests out of nine successful (56%). Since two of the three nests of unknown outcome made it at least to the nestling stage, actual nest success was probably closer to 56% than to 22%, as most condor nests observed at the nestling stage have resulted in fledged young (88%).

For the DDT years, 30 condor nestings were documented, but 18 of these must be excluded from analysis because they were nests found at the nestling or fledgling stage that probably would not have been found if they had failed at the egg stage. Of the remaining 12 nests, 10 were followed to a known conclusion, while two had unknown outcomes. A majority of the analyzable nests were found relatively late in the breeding season but were located in sites being checked regularly for nesting activity so that there is a reasonably good chance that early failures would have been detected. Minimum nest success for the DDT years was four nests out of 12 successful (33%); maximum success was six nests out of 12 successful (50%).

For the post-DDT years, 22 nestings have been documented, but 17 of these must be excluded from analysis either because they were found at the nestling or fledgling stage and would not have been found if they had failed at the egg stage or because they failed due to human disturbance. Of the remaining five nests, two were successful (40%).

In many bird species the primary negative effect of DDT contamination has been eggshell thinning, resulting in increased egg breakage and reduced nest success (*see* Cooke 1973, Stickel 1975). Significant DDT contamination and eggshell thinning have been documented for the California condor during the DDT era by Kiff et al. (1979). The condor nesting data, however, do not provide support for large impacts of DDT on reproductive success. In all three eras considered above, nest success approached 50%, and there are no significant differences in nest success between the DDT and non-DDT years. Similarly, the apparent rate of egg breakage, although high, is not clearly different in the DDT and non-DDT years. For the pre-DDT years three (possibly four) of nine analyzable nests suffered apparent egg breakage (33–44%), while five (possibly six) of 12 analyzable nests during the DDT years suffered apparent egg breakage (42–50%). For the post-DDT years none of five analyzable nests is definitely known to have suffered egg breakage (0%), although two nests could have failed to this cause.

Further support for a naturally high rate of egg breakage in the species, independent of the effects of DDT, comes from data in egg collections. According to Lloyd Kiff of the Western Foundation of Vertebrate Zoology (pers. comm.) approximately 10–15% of the condor eggs collected pre-DDT were already punctured or otherwise damaged when they were first found. Since collectors have discovered many nests by observing the activities of adult birds and since adults cannot be expected to long remain in attendance at sites with thoroughly broken eggs, the chances of collectors discovering thoroughly broken eggs might have been biased on the low side. In addition, it is likely that some collectors happening upon remains of broken eggs may have failed to note or collect them because they were of no particular value to the collectors. These factors could result in the frequency of damaged eggs in collections being an underestimate of the true natural breakage rate. However, the situation is clouded by the fact that to some extent collectors have increased the damage rate by occasionally causing incubating birds to flush from their nests in a reckless manner.

Conceivably the absence of obvious increases in rates of egg breakage and reproductive failure during the DDT era may trace largely to the small sample sizes of analyzable nests and to the swamping of DDT effects by high natural breakage rates. It is difficult to believe that the amount of eggshell thinning documented in condors for the 1960s (averaging over 30% thinning) might have been completely harmless, as in other species shell- thinning of even 20% has been associated with severe declines in reproductive success. Nevertheless, it is worth pointing out that a number of condor eggs with thinning close to 30% have hatched successfully, and that respectable numbers of fledglings were documented during the DDT era. It is possible that condors are relatively resistant to the penalties of shell-thinning.

Whatever the impacts of DDT may have been in the 1960s, it appears that eggshell thickness in the species has since been recovering to more normal levels (Snyder et al., in prep.). Shells from three nests (two pairs) of 1980–81 average only 1 to 9% thinner than normal, although a third pair active (and successful) in 1981 has been laying eggs with shells close to 30% thin (Figure 2).

Excluding failures that were definitely human-caused, most nest failures have occurred at the egg state, with clearly or possibly broken eggs found in ten nestings from 1939 to the present. Of the four instances from 1939–1946, three were detected during the normal incubation season. In contrast, of six instances from 1947–1971, all were detected after the normal incubation season. Especially for these latter (DDT) years, some of the apparent egg breakage may really have been egg destruction by common ravens (*Corvus corax*) or terrestrial mammals (possibly in some instances after eggs were deserted because of infertility or addling). In addition, some breakage may have resulted from a natural clumsiness of condors in manipulating eggs and a tendency for some pairs to nest in caves with rocky bottoms. Six of the ten apparently broken eggs discovered between 1939 and 1971 came from just two nests holes (S41 and S231). For each of these holes breakage occurred during both the pre-DDT years and the DDT years. S41, which accounted for four instances of possible breakage, is one of the most cramped condor nest holes known, and had a very rocky floor until 1967, when Fred Sibley cleared out the rocks on the assumption that frequent egg breakage might have been due to the roughness of the floor. The constricted dimensions of this hole might have made it difficult for the adults to turn around and position their eggs carefully. Direct observations of one nest in 1980 (S353) have made it clear that even in spacious sites, condors have physical problems in manipulating their eggs; the adults at this nest were sometimes observed kicking their egg across the nest floor while attempting to settle on it.

The sample size of condor nests analyzable for nest success is small, with some specific nest sites and pairs represented more than once. The overrepresentation of some pairs and sites affects the confidence with which one can project results to the species as a whole. By historical accident, a significant fraction of the nesting attempts that qualify for analysis (4 of 25) took place in the nest hole discussed above in which four apparent instances of egg breakage took place. Assuming that the breakage may have been at least partially caused by poor quality of the site, it is possible that the calculated egg-breakage and nest-failure rates are biased upward as a result. If this site is excluded from analysis, egg-breakage rates drop from 33–44% to 25–38% for the pre-DDT years, and from 42–50% to 33% for the DDT years. On the same basis, nest success rises from 22–56% to 25–63% for the pre-DDT years, and from 33–50% to 44–56% for the DDT years. The latter rates may be closer than the former to true rates for the species.

ADEQUACY OF REPRODUCTION

Whether the decline of the California condor can be largely or in part attributed to reproductive difficulties is one of the central questions confronting the present research program. In the preceding section I considered reproductive success of pairs from 1939 to the present and found that roughly 50% of the analyzable nests resulting in fledglings. As this figure closely resembles the nest success of a variety of raptor species (*see* Brown and Amadon 1968), it does not in itself appear to be low enough to be clearly a major cause of the population decline. It is possible that 50% nest success may be "normal" for the species, and the primary causes of decline may lie elsewhere in the life cycle—perhaps in a failure of significant numbers of adults to attempt breeding, in poor survival of birds from fledging age to independence, or in poor survival of birds after independence. However, the high and largely unexplained rate of egg breakage demands further attention, and it is conceivable that nest success should be, and perhaps once was, considerably higher than 50%. Since the species has been declining, we must assume that one or more components of reproduction and survival have been suboptimal. The question is which ones. Unfortunately, nothing comprehensive is known about survival of any age class of condors past fledgling, and nothing direct has ever been determined about the percentage of potential breeding birds actually breeding. Thus, it has been difficult to get any reliable indications as to where the primary problems have lain.

Although the dearth of information on condor survival and reproduction has been discouraging, recent field observations have made possible a narrowing of the range of potential difficulties. During 1981, sightings by the condor research center staff and their cooperators have revealed the apparent existence of eight condors with completely dark heads and necks in the wild population (not including nestlings of 1981). On march 31, 1981, five dark-heads were seen at close range on a ranch in the southern San Joaquin valley. In addition, we can now account with reasonable confidence for three other dark-heads in existence at the same time, distinguishable from the above five birds by characteristics of behavior and plumage (two of them dependent fledglings in the spring of 1981 and the other a visually distinctive bird).

The presence of eight dark-headed condors in the population is significant because of what can be inferred with regard to how many pairs have been breeding in recent years. From the information available on maturation of color characteristics of condors, it appears that juveniles retain dark heads until they are about 3–4 years old (*see* Koford 1953, Todd 1974, Wilbur 1975). A captive of known age studied by W. L. Finley early in the century had a completley dark head until it was at least 2 years 8 months old. Another captive held at the National Zoological Park early in the century had an orange head by the time it reached an age of four years. Two condors held at the Selig and San Diego zoos were reported to undergo a head color change from gray to orange between four and six years of age. Another captive that died at five years of age in the Selig Zoo was nearly adult in characteristics. TopaTopa, the captive condor currently at the Los Angeles Zoo, had the orange neck of a "ring-necked" bird in the winter of his third year and began to develop orange head color at almost exactly four years (in May). These data are few and may be biased because they are based entirely on captives. However, if we assume that development proceeds at roughly the same rate in the wild as in captivity, we can tentatively conclude that condors seen in the spring with completely dark heads and necks are probably no more than about three years old. Thus, the most reasonable conclusion is that the eight dark-heads seen during 1981 indicate a minimum of eight condors fledged 1978–1980.

How many pairs did it take to produce this many dark-heads? Available evidence suggests that it generally takes two years for a condor pair to rear a single juvenile to independence, and that a pair producing surviving fledglings usually lays only a single egg every other year. In observations of 1980–81, adults with dependent fledglings during the second spring of the breeding cycle commonly showed signs of renewed interest in breeding (including checking of potential nest caves) but did not follow through with egglaying. Koford (1953 and field notes) also observed pairs that were apparently inhibited from starting new breeding cycles through the spring months by presence of dependent fledglings. While evidence that some pairs might fledge young in consecutive years has been obtained in several cases (*see* Sibley 1970; Jan Hamber, pers. comm.), in none of these, with one questionable exception, has a fledgling been documented in association with its parents as they have started a new breeding cycle. Quite possibly in some or all of these cases, fledglings have not survived their first winters, although early independence and dispersal of fledglings cannot be ruled out conclusively.

If we assume that biennial breeding of successful pairs is the rule, a single pair's maximum production of young that might still be dark-headed through the spring of 1981 would be two—one produced in 1978 and one in 1980. On the other hand, if the pair bred on the alternate year scheduled—i.e. 1977, 1979, etc.—the maximum number of young still completely dark-headed through the spring of 1981 would be only one. Hence, if all pairs laid eggs in 1978 and 1980, if all were successful in breeding and if no fledgling mortality occurred in the past three years, a confirmed minimum of eight dark-headed juveniles would imply a minimum of four breeding pairs. In fact, however, it is clear from direct observations that not all breeding attempts in the past three years have been successful, and not all pairs have been breeding on a 1978–1980 schedule. It is also reasonable to guess that survival of fledg-

Table 4. Known and strongly suspected production of young by 5 California condor pairs believed active, 1978–1981

Pair[a]	Nesting activities during indicated year[b]			
	1978	1979	1980	1981
1	?	?	N-F	F
2	?	N-F	F	X
3	?	?	0(?)	N-F
4	?	?	N-F	F
5	0(?)	0(?)	0	0

[a]Most of these pairs (or pairs they have replaced in the same nesting areas) have been reproducing, at least intermittently, through most of the 1970s. Fledglings were documented for pair 1 in 1972 and 1975; for pair 2 in 1976; for pair 3 in 1975 and 1976; for pair 4 in 1968; and for pair 5 in 1972, 1974, 1976, and 1977. Additional undocumented fledglings may have been produced in these areas during the 1970s.
[b]N = Nestling; F = Fledgling; 0 = Nest failure; X = One member of pair disappears.

lings probably has not been 100% during the same period. Furthermore, there is direct evidence of five breeding pairs probably active during this period.

The known and suspected breeding histories of the five pairs believed active since 1978 are given in Table 4. These pairs produced three known (or strongly suspected) young during 1978–1980. Assuming biennial breeding for successful pairs, they could have produced a maximum of six (or at the very most seven) young. Thus it is reasonable to postulate the existence of at least one more pair in the wild population to account for the eight dark-headed juveniles of 1981 (i.e. six pairs).

The theoretical number of young a pair can be expected to produce over a three-year period depends on nesting success, the interval between successful breedings, and the probability of renesting within a breeding season in the event of early nesting failure. If we assume no renesting, 50% nest success, and biennial breeding for successful pairs, the number of young expected per pair would be very close to one in a three-year period. If we assume 50% of the nesting failures might occur early enough that a renesting attempt might take place in the same year, the expected number of young over a three-year period would increase to slightly more than 1.3. There is no comprehensive information on frequency of renesting following failure within breeding seasons, but I believe it is reasonable to speculate that 50% might not be too far from the mark. Good evidence of renesting in condors has been obtained in only two instances (Harrison and Kiff 1980, and observations of Jan Hamber in 1981), but the absence of other documented examples could easily be an artifact of a strong tendency for condor pairs to switch nest sites after failure and a difficulty in following pairs after failure. The two nest sites apparently used by a single pair in 1981 were situated about 10 km apart, and other pairs of recent years have been observed investigating alternative nest holes separated by comparable distances.

Thus, we might expect six pairs to be capable of producing as many as eight young over a three-year period, given normal nest success and biennial breeding for successful pairs. If we now include a reasonable estimate of 10–15% annual mortality for juveniles, it would take seven or eight pairs to produce eight surviving young. These calculations suggest that we still have not located several of the breeding pairs recently active in the wild population.

Further, if one assumes the existence of seven or eight breeding pairs in 1978–1980, and to this total adds a minimum of eight dark-headed juveniles and a minimum of two subadults believed to exist in 1981, one accounts for 24 to 26 birds in the population during this period, a total very similar to the central population estimate of 25 we have been using, and only slightly less than Wilbur's population estimate of 30 for 1978. That the calculated total comes out so close to earlier estimates without any allowance for non-breeding adults in the population, suggests that most adult condors have been attempting to breed in recent years. Actually 30 birds is probably a better estimate of the 1978–80 population size than 24–26 birds, when one considers the near certainty that the sex ratio in the adult population may deviate to some extent from equality and the possibility that we still have not accounted for all juveniles and subadults in the wild population.

An apparent minimum of eight dark-heads in the wild suggests a recent production of three or more fledglings per year. Using a range of mortality estimates for juveniles and adults, Verner (1978) calculated for a population size of about 50 condors that necessary annual production of young would have to be about five or six to achieve population stability. Scaling this estimate down to a population of about 30 birds, one might expect necessary production to lie in the range of 3 to 3.6 young per year. The recent production of fledglings apparently lies within this range. However, it must be emphasized that true mortality rates in the wild population are unknown. If they are greater than the range postulated by Verner, the numbers of young needed annually for population stability should be modified accordingly. For the present, the adequacy of recent reproduction can only be assessed in conjunction with adequacy of survival rates, and only by monitoring overall population trends. As yet there have been no clear indications from counts in high condor use areas that the population decline might be slowing or re-

Table 5. Number of immature condors reported for wild populations by Wilbur (1978)

Year	Number of immature condors
1968	10
1969	10
1970	13
1971	8
1972	8
1973	6
1974	4
1975	5
1976	6

versing itself, so it is likely that problems still exist with either surival rates or some aspects of reproduction, or both.

The number of immatures believed to exist in 1981 (10, including subadults) is considerably higher than the numbers presented by Wilbur (1978) for other recent years (Table 5), a result that could be interpreted to indicate a recent increase in reproduction. However, it seems more likely that the number of juveniles found, like the number of nests found in the past two years, represents only an apparent increase. As discussed earlier, the number of field-days expended by condor researchers has increased greatly in the last two years. Further, the chances of confirming the existence of large numbers of juveniles must be a function of observer effort. The truth of this assertion is especially apparent for the 1981 data, where the spring accounting for ten immatures (including two subadults) depended heavily on a single fortuitous observation of five dark-headed birds seen together. In the absence of this sighting and a number of other observations clearly attributable to increased field coverage, the estimated number of immatures known in the population would probably not have exceeded 5 or 6.

DISCUSSION

Most of the conclusions reached in this paper are tentative, as they are based on very small sample sizes of observations and on extrapolations of uncertain rigor. Nevertheless, I feel it is important to present these conclusions, speculative though they are, because they at least help focus future research and conservation efforts. With a species as difficult to study and declining as rapidly as the California condor, it may not be possible to achieve a thorough understanding of the causes of decline before extinction occurs, and it is crucial to continuously re-evaluate the data that do exist.

Available data suggest a recent natural nests success rate of about 50% for the California condor, a rate that is not necessarily any lower than what the species has enjoyed for millenia. This is not to say that the condor would not benefit from efforts to increase nest success. Most failures have occurred at the egg stage, with apparent egg breakage occurring in a large percentage of cases. Some of the breakage may be due to poor quality of the nest sites and some may be due to the action of natural predators. There are potentials for reducing both these sorts of stresses. Unfortunately, the major causes of egg breakage are still unknown. Truly effective efforts to increase nest success will depend on elucidation of these causes.

The apparent minimum of eight dark-headed condors documented in 1981 indicates greater reproductive effort in the wild population than had been observed in immediately preceding years, but this should not be construed as evidence for true increases in reproduction of for recovery of the population, since the intensity of field observations in the past two years has been running about five times as great as the average of immediately preceding years. Just how large a breeding population is implied by the dark-headed birds depends critically on a number of assumptions. I have developed an estimate of seven or eight breeding pairs assuming that the eight dark-heads observed in 1981 were all three years old or less, that nest success has been about average in the past few years, that about half the nesting failures have been followed by renesting attempts in the same breeding season, that successful pairs have bred on a biennial basis, and that juvenile mortality has run about 10–15% annually. But if some of the dark-heads were really cryptic ring-necks (ring-necks are sometimes very difficult to distinguish from dark-heads in the field), if some wild juveniles retain dark heads and necks through their fourth spring, if successful pairs nest on an annual basis with some frequency, or if by chance nest success and juvenile survival have been greater than usual in recent years, seven or eight breeding pairs could be an overestimate. Clearly, more comprehensive information is needed on the numbers of breeding pairs, breeding frequency, nest success, juvenile survival, and ages of dark-headed juveniles.

While recent data have weakened the hypothesis that reproduction has been a significant problem for the condor, a rigorous evaluation of condor reproduction is still not possible. If preliminary extrapolations are correct, we still have yet to locate the nesting areas of several breeding pairs in the population. Finding these hypothetical pairs promises to be very difficult if techniques continued to be limited to traditional naturalistic methods. Similarly, no practical methods for evaluating the impact of various mortality factors on the condor are available to the research program, and there is, in fact, no good information on age-specific mortality rates. Chances are slim that a comprehensive evaluation of either mortality factors or reproduction can be achieved in the absence of extensive radio-telemetry studies.

SUMMARY

Although more condor reproduction has been documented in the past two years than in immediately preceding years, the increase is likely only an apparent one, tracing primarily to greatly expanded observer effort.

Overall nesting success of condors has averaged about 50%, and this rate has not changed noticeably since 1939. However, most nests have unfortunately been located late in the breeding cycle, and the sample size of nests useful for nest success calculations has been relatively small, with some pairs overrepresented, leaving considerable uncertainty about true values of nesting success over the years.

Most nesting failures have occurred at the egg stage, with apparent egg breakage reported in a surprisingly large proportion of the nests. Causes of the breakage are largely unknown, but factors other than DDT contamination appear to be dominant, as frequent egg breakage has occurred throughout the known history of the condor and breakage rates did not increase dramatically during the DDT years. That DDT contamination might have been a major factor in the recent decline of the species is not obvious from the nest success and egg breakage data.

Whether reproductive problems have been a major cause of decline of the species is still unclear. Recent nest success has not been so low as to indicate severe difficulties, and recent counts of free-flying juveniles in conjunction with total population estimates imply that most adults may be breeding. These findings suggest in a preliminary way that the primary problems of the species may prove to be ones of mortality rather than reproduction, but much more information is needed before this conclusion can be considered firm.

Recent sightings of juveniles suggest that the number of condors in the wild population from 1978 to 1980 might have been about 30, in agreement with the estimate of Wilbur (1980), although this should be considered a very rough estimate. The number of condors present today could well be lower.

ACKNOWLEDGMENTS

The present California condor research program is a cooperative effort deriving support from the U.S. Fish and Wildlife Service, the National Audubon Society, the U.S. Forest Service, the Bureau of Land Management, the California Department of Fish and Game, and numerous other contributors. Primary research responsibilities lie with the U.S. Fish and Wildlife Service and the National Audubon Society. The results reported in this paper rest upon data collected by a great many individuals over a span of several decades. Special thanks to John C. Ogden and Jared Verner for critical comments on various of the ideas presented.

LITERATURE CITED

Borneman, J. C. 1965–1980. Unpublished field notes, condor research center.
Brown, L., and D. Amadon. 1968. Eagles, Hawks and Falcons of the World. McGraw-Hill, New York.
Cooke, A. S. 1973. Shell thinning in avian eggs by environmental pollutants. Environ. Pollut. 4:85–152.
Harrison, E. N., and L. F. Kiff. 1980. Apparent replacement clutch laid by wild California condor. Condor 82:351–352.
Kiff, L. F., Peakall, D. B., and S. R. Wilbur. 1979. Recent changes in California condor eggshells. Condor 81:166–172.
Koford, C. B. 1953. The California condor. National Audubon Soceity Research Report 4.
Miller, A. M., McMillan, I., and E. McMillan. 1965. The current status and welfare of the California condor. National Audubon Society Research Report 6.
Ogden, J. C., and N. F. R. Snyder. 1981. The view from Ventura. Point Reyes Bird Observatory Newsletter 53:11–15.
Sibley, F. C. 1966–1969. Unpublished field notes, Condor Research Center.
Sibley, F. C. 1970. Annual nesting of the California condor. Paper presented at annual meeting, Cooper Ornithological Society, Fort Collins, Colorado, 20 June 1970. 3 pp.
Stickel, W. M. 1975. Some effects of pollutants in terrestrial ecosystems. pp. 25–74. in Ecological Toxicology Research, A. D. McIntyre and C. F. Mills (eds.), Plenum Publ. Corp., New York.

Todd, F. S. 1974. Maturation and behavior of the California condor at the Los Angeles Zoo. Int. Zoo Yearbook 14:145–147.

Verner, J. 1978. California condor: status of the recovery effort. U.S. Forest Service General Technical Report PSW–28, 1978.

Wilbur, S. R. 1975. California condor plumage and molt as field study aids. Calif. Fish Game 61(3):144–148.

Wilbur, S. R. 1978. The California condor, 1966–76: a look at its past and future. North American Fauna 72.

Wilbur, S. R. 1980. Estimating the size and trend of the California condor population, 1965–1978. Calif. Fish Game 66(1):40–48.

The California Condor Recovery Program: An Overview

John C. Ogden

FIELD studies and surveys of the California condor (*Gymnogyps californianus*) conducted at intervals between 1939 and 1979 by Carl Koford, Fred Sibley, Ian and Eben McMillian, Sanford Wilbur and others have revealed important aspects of the biology, distribution, status and history of this severely endangered species (Koford 1953, Miller et al., 1965, Wilbur 1974, 1978, 1980). These studies have suggested a number of specific courses of action for protecting condors, primarily through programs designed to isolate condors from direct human disturbance and to increase protection of their habitat (Wilbur et al. 1980, Carrier 1971). Although several Federal, State and private agencies and organizations have been generally successful in implementing a series of such protection programs, the condor has continued to decline at an alarming rate. The most recent analysis of annual census data indicates that the total number of condors dropped from an estimated 50 to 60 birds in the late 1960's, to between 25 and 35 condors by 1978 (Wilbur 1980).

The growing conviction that the California condor is rapidly moving towards extinction has recently prompted several independent reviews of the condor's situation (Ricklefs 1978, Verner 1978, Clark 1979). Although the recommendations by the review groups have differed in detail, they agree on two main points: (1) that isolating condors and their habitat from human contact as a means for stopping their decline, even where these efforts have successfully protected nesting and roosting sites, has been inadequate as a means for stabilizing the wild population, and (2) that new, more active programs for the preservation of the California condor should be considered. A major recommendation of the American Ornithologists' Union report (Ricklefs 1978, p. 15) was that the situation is so critical and that the factors that are causing the recent decline are so poorly known "that the only hope for the species lies in a long-term, large-scale program involving greatly increased research effort, immediate steps to identify and conserve vast areas of suitable condor habitat, and captive propagation."

These recommendations were similar to those contained in a Contingency Plan and its subsequent revision, prepared by the California Condor Recovery Team. (Wilbur et al. 1977, Eno et al. 1979). The Contingency Plan, in the event that implementation of the formal recovery plan did not halt the decline, called for the addition of a captive breeding program and increased research on the wild population.

The Contingency Plan and the A.O.U. report became the basis for a greatly intensified recovery effort for the California condor, which is discussed in the following pages. The intensified recovery effort was formally initiated with the signing of a cooperative agreement during December 1979 by the U.S. Fish and Wildlife Service, the National Audubon Society, California Department of Fish and Game, U.S. Forest Service and the Bureau of Land Management. The major initial components of the new recovery program put in motion by the cooperative agreement are a radio-telemetry study and a captive breeding program, with lead responsibility for these tasks in the hands of the U.S. Fish and Wildlife Service and the National Audubon Society. The lead organizations each agreed to provide one senior biologist, Noel F. R. Snyder for the Service and John C. Ogden for Audubon, with the two having joint responsibility for directing the telemetry and other field research programs. Responsibility for captive breeding is being handled through cooperative programs between the U.S. Fish and Wildlife Service, the California Department of Fish and Game, and zoos in San Diego and Los Angeles. The decision to start the intensified recovery effort with concurrent telemetry and captive breeding programs is based upon the following analysis.

A basic goal of the recovery program is to maintain a population of condors in their native range if at all possible. Ideally this goal should be accomplished by reversing the present decline before the wild population disappears, rather than by depending solely on the captive breeding program for its maintenance. The rate of decline derived from Wilbur's (1978) estimate of numbers of condors, however, if continued in a linear fashion, will see the end of the wild population in only 12 to

15 years. With no evidence to indicate that the status of the condor has improved since Wilbur's analysis, it is essential that the intensified recovery program contain the option to utilize those research and management techniques that are most likely to both identify and correct the adverse factors responsible for the condor's decline within this 12- to 15-year time period.

The need for telemetry as the best tool available to the researchers for resolving key questions about the condor and its habitat is derived from the fact that traditional field techniques of watching unmarked condors from great distances, although tried for 40 years, have failed to produce the kinds of information upon which to base a successful recovery program. With so little time now potentially remaining for the condor, telemetry is the only field technique with the capability of providing the kinds of detailed daily activity and location information on condors that is needed to conclusively determine habitat needs and the factors responsible for the decline. Although the telemetry program offers the best chance of obtaining significant amounts of new biological information on the condor, just how long before telemetry will provide this new information remains unknown.

Once the key factors responsible for the decline are identified, there is still a question of whether these factors can be corrected before the wild population disappears. Thus, the need for a concurrent captive breeding program is by far the best insurance against the uncertainties of the telemetry and protection programs. Condors produced by a breeding program can be successfully released into the wild if the results of the surrogate work with Andean condors are indicative of how captive-bred California condors will behave in the wild (S. Temple and M. Wallace, pers. comm.). Captive-bred condors can also be used to increase the size of the wild population at a much greater pace than can possibly occur through natural reproduction, as well as for re-establishing condors in portions of their range where they have been extirpated.

The goals we seek to achieve with the aid of telemetry include the following:

1. Identify the habitats and specific sites that are important to the present population of wild condors, with special attention to feeding sites, and the use of lands in both public and private ownership.

2. Locate all nesting pairs.

3. Determine causes of nesting failure and causes of failure to attempt breeding.

4. Determine adequacy of food resources.

5. Measure daily and seasonal movements of condors related to feeding and nesting activities.

6. Learn causes of injury or mortality to condors.
7. Improve censusing of the wild population.

The telemetry program will allow for collection of daily locations and movements for individual condors, and thus will provide the capability to begin to answer a host of key questions related to the above list of topics. Heretofore, the vast and rugged nature of condor habitat, and the wide ranging behavior of the birds have made it impossible to locate all active nests, to answer almost any questions regarding daily and seasonal movements of individual birds related to environmental influences, or to accurately census the size and age structure of the wild population. Efforts to deal with these questions through the use of co-ordinated surveys involving groups of strategically locate observers, even where 2-way radio communications between observers were utilized, have generally resulted in disappointing and limited results. Annual attempts to determine the size of the wild population by these naturalistic techniques have been inconclusive, with each survey subject to a variety of interpretations depending on how birds that were seen at different stations are matched up, and what kinds of assumptions are made about the distances that condors may move in a day (Wilbur 1980). After two years of intensified field study of condors, without the use of telemetry, we are more convinced than ever that the rate of acquisition of new information has been far short of what may be needed to save the species.

Key elements of the trapping and telemetry program are as follows. Condors will be trapped with cannon nets, a capture technique that was decided upon following a thorough review and testing of a variety of trapping devices. Rocket nets, bow nets, clap traps, drop traps, walk-in traps, padded jaw traps and cannon nets were evaluated, and those techniques that appeared most promising (cannon and rocket nets, clap trap and walk-in trap) have been field tested by the California condor research team on Andean condors in Peru, and on several species of both Old and New World vultures in Zimbabwe, Africa, and in California. The cannon net, based in part upon its good performance when used by members of the Vulture Study Group in Southern Africa for capturing over 900 vultures (P. Mundy, pers. comm.) and on its proven high efficiency, selectivity, simplicity of operation and extremely low probability of causing injury to either birds or operators, in our judgment is the superior trapping procedure. The similar rocket net, used with such success with Andean condors in Peru (S. Temple and M. Wallace, pers. comm.) with many of the same admirable features as the cannon net, is a good back-up system.

Each condor for the telemetry program will be radioed with two wing-mounted (patagial) transmitters, one on each wing. Patagial mounting of the transmitters is superior to other attachment schemes in that it combines the advantages of a relatively simple, light-weight attaching procedure (a small vinyl tag), provides for the acquisition of a long-term history from each bird due to the relatively permanent attachment, without the need for frequent retrapping, and is ideally situated on the condor for exposure of solar panels to full light. Each transmitter package, including the attachment tag that supports the radio, will weigh between 40 and 45 grams. The solar-powered, primary transmitter, designed by Dr. William Cochran at the University of Illinois Natural History Survey, has an estimated life of two to four years, and a maximum signal range of 75 to 100 miles. The second unit is eventually expected to be either lithium battery powered, or have a combination battery and solar power source. This back-up unit will serve both a secondary function for general tracking and as the primary trackable unit in the event that a condor dies or is injured and the solar unit becomes covered or inverted.

Condors will be tracked in three ways. Birds will be followed by mobile ground teams using hand-held receivers, by receivers carried in a Cessna 180 aircraft with directional antenna mounted beneath the fuselage, and through a network of automated receiver stations located on selected mountain-top towers. The latter tower system has been developed by Cochran, and will consist of a number of fixed receiver stations each relaying condor position data collected at predetermined intervals through either telephone or microwave systems to a central computer located at the Condor Research Center. The location data from each condor will be plotted and mapped, and will become the basis for the more intensive field studies of individual birds in an effort to answer the questions noted above.

The program to capture California condors for telemetry and captive breeding is a major departure from earlier, more naturalistic field studies of the condor, and is the key aspect of the intensified recovery program that has prompted a vocal minority of conservationists and scientists to offer opposition (*see* Holden 1980, Kahn 1980). The fact that the intensified condor recovery program has become so controversial is worth exploring, not only because the content and pace of the condor program have apparently been altered as a result of the controversy, but also because the bio-political issues raised here include many of the same questions that have plagued earlier programs to save other endangered species. A review of these issues may be useful to those persons who have a role in future recovery programs.

The questions over the propiety of the proposed condor programs have occurred at two levels. First has been the question of whether the effort to save the condor should be made at all. A small number of people feel that the effort to save the condor should not be made, either because the cause is hopeless or because other conservation issues should have greater priority (Stallcup 1981, Pitelka 1981). The large sums of money being spent for the condor, they argue, could better be used for some other conservation effort. In the case of the condor, however, this rather philosophical question was addressed during the late 1970s, when the Fish and Wildlife Service, the National Audubon Society, and other organizations reviewed the existing data and recommendations on the condor, and concluded from these data that the effort to save the bird should be made. Considering that the factors responsible for the condor's decline are not precisely known, and that major habitat questions remain to be answered, it is premature at this time to be adopting a doomsday attitude about the condor's future. The argument that money spent for the condor might better be used for some other environmental issue assumes that such money would still be available for environmental programs if not used for condors. In fact, much of the condor money is coming from a special congressional appropriation to the Fish and Wildlife Service earmarked specifically for condor recovery, and from special contributions to the National Audubon Society that were identified for use on the condor program. These monies became available as a direct result of a broad-base of support from the conservation and ornithological communities for an accelerated effort to save the California condor (*see* Resolutions of the American Ornithologists' Union annual meeting, 1980, and the Cooper Ornithological Society annual meeting, 1981). It is not reasonable to expect that these funds can be readily transferred to other environmental programs if not used for condors.

Once that decision is made, then the second question becomes one of how the condor should be saved, and here is the basis for the major opposition to the proposed breeding and telemetry programs. A central feature of the controversy has been the lack of agreement over which factors in the condor's environment are responsible for the continued decline. The results from earlier condor studies leave little or no doubt that condors in the past have been shot, have been the unintentional victims of predator and/or rodent poisoning programs, apparently suffered egg-shell thinning during the DDT era, have been disturbed at nests and roosts by everyone from egg collectors and photographers to picnickers and hikers, and have lost important portions of their habitat (Koford 1953, Miller et al. 1965, Wilbur 1978, Kiff et al. 1979). Opponents of the proposed telemetry and captive breeding program claim that one or more of these past factors is still responsible for the continuing decline. In this list, habitat loss is the one factor most often suggested as responsible for the condor's plight. With their strong convictions as to what the condor's problems are, the critics argue that the expanded research may be unnecessary, and that the breeding program is premature. Instead, they say, we should be devoting most of our energy to preserving habitat, reducing shooting and poisoning and protecting nest sites. Various conservation organizations, including the Friends of the Earth, elements of the Sierra Club and a few California chapters of the Audubon Society, have been among the major proponents of these views. The more extreme among the opponents argue that almost any course of action is preferable to trapping condors, as the latter action in their opinion will surely hasten the condor towards extinction (Philips and Nash 1981). This latter opinion is based more on emotional and philosophical considerations than on any knowledge of wildlife research or trapping techniques.

While we readily agree that some or all of the factors mentioned above have operated in the past to reduce condor numbers, we see little in the way of compelling evidence to indicate which, if any, of these factors may be operating today, or whether these factors are more responsible for the recent decline than some others that are not presently being considered. In fact, even though biologists and bird-watchers have spent many thousands of observer-days in the field studying and searching for condors in recent years, there have been no confirmed reports of a condor shot since 1976, or of a probable poisoned bird since 1966 (Wilbur 1978, recent unpublished field notes). At this point, our critics are asking us to initiate greatly intensified management and protection efforts without knowing what the bird's present problems are, and gambling that we address the correct adverse factors in the potentially short time remaining for the condor. In our view, such a request is unacceptable.

In the case of questions about habitat, it is yet to be demonstrated that the remaining condor population is in any way stressed due to a lack of suitable habitat. California condors utilize two basic types of habitat, the rugged chaparral and pine-covered mountains where they nest, and the open grasslands where they do most of their feeding. Vast amounts of the former habitat types still exist within National Forests in Southern California, including the locations of all condor nests known to have been active for many years. The U.S. Forest Service has taken vigorous and effective steps to protect known condor nesting and roosting sites under their jurisdiction, and it seems most unlikely that lack of sufficient nesting sites or excessive disturbances of these sites is an important factor in the recent decline.

If condors are experiencing any serious habitat-related problems, they are probably occurring on the grassland feeding grounds in private ownership. Today's condors apparently feed to a large extent on carcasses of domestic livestock and native mammals that they find on rangelands around the southern borders of the San Joaquin Valley. Our aerial surveys of this region, over four large counties, reveal that great expanses of rangeland that are relatively isolated from frequent human disturbance still remain in the condor's range. A detailed survey of land-use patterns in one of these counties (Cynthia Studer, in prep.) has shown that while there has been a trend towards gradual reduction in overall acreage of rangeland and number of head of livestock, no large-scale reductions in either have occurred during the past decade. If the trend in this one county is representative, then it seems unlikely that enough condor feeding habitat has been lost since the late 1960s to explain the decline in condors reported by Wilbur (1978) for this period. Further support for this view is the fact that the two pairs of nesting condors that were closely monitored by our staff, one in 1980 and a different pair in 1981, had no apparent problem finding sufficient food for their chick (*see* further discussions in this paper, and Table 2).

Our present understanding of the precise location of important condor habitat, while fairly good for nesting and roosting sites on the National Forests, is quite poor for the feeding habitat on private lands. We are in no position to recommend allocations of money or law enforcement for the protection of specific parcels of privately-owned lands until we have a better understanding of where condors are feeding, and whether human activities on those lands may be causing problems. At present, in spite of our best efforts, the location of at least half of the remaining condors remains unknown for many months of the year.

MAJOR EVENTS OF 1980–82

Political and Regulatory Matters

Capture of California condors requires prior issuance of a Federal permit from the Office of Endangered Species, U.S. Fish and Wildlife Service, and a State permit from the California Fish and Game Commission. Each permitting process includes a requirement for a public review on the permit application prior to its issuance. For such a controversial program as the proposed trapping of California condors the review period included a series of public meetings whereby organizations and individuals were given the opportunity to question and discuss the recommended condor recovery program. In addition to the rather vigorous suggestions that habitat protection rather than the "hands on" trapping program should be the priority, other concerns were expressed. These included the opinions that such a controversial program requires preparation of a federal Environmental Impact Statement rather than the different level of analysis characteristic of the Environmental Assessment prepared by the Fish and Wildlife Service, that need for the proposed telemetry and breeding programs is not supported by proof of the severe nature of the condor's status, and that the proposed trapping, telemetry and captive-breeding protocols have not been adequately tested on surrogate species. As we eventually learned, these criticisms actually were a screen to mask the more fundamental and philosophical objections to the recovery program by those groups or individuals who opposed any trapping of condors. The more deep-rooted basis for much of the program's opposition became increasingly apparent as the Fish and Wildlife Service and National Audubon Society responded to the above criticisms by adequately explaining the legality and adequacy of the Environmental Assessment, presented thorough documentation of the deteriorating status of the condor and the need for prompt action as the best approach for saving the species, and documented the considerable extent to which the proposed trapping, telemetry and breeding procedures have been tested and evaluated. These data did not greatly alter the positions or arguments by those in opposition since, in truth, they were less interested in the biological and legal questions than they were in slowing, stopping or altering the recovery program by creating the illusion that a major split existed within the environmental and scientific communities over the questions of how the condor could be saved. The creation of such an air of controversy was designed to influence the state's permit-issuing body, the Fish and Game Commission, as well as the Department of Fish and Game, to proceed with considerable caution in reviewing and authorizing the recovery program.

Following public review, both Federal and California permits were approved during May 1980, authorizing up to ten condors for radio telemetry, and one condor for captive breeding during the first year of the program. One month later, however, following the death of a nestling condor while it was being weighed and measured as part of a study of growth rates, the state permit was withdrawn. Opponents to the condor program took this opportunity to renew their efforts to either halt the program or change its priorities. In addition, the California Fish and

Game Commission organized a Condor Advisory Committee chaired by an outspoken critic of the telemetry and breeding programs to review and make recommendations to the Commission on future condor recovery efforts. Thus, the state's advisory committee became the fifth advisory or review committee looking at the California condor recovery program (the other four: Fish and Wildlife Service, California Non-Game Advisory Committee, Sierra Club, and American Ornithologists' Union). Interestingly, and as might be expected from groups with such diverse backgrounds and philosophies, the advisory committees have not often agreed on the priorities and content of the condor program.

As good an example as any of the disagreement between advisory committees was the 1980 A.O.U. Committee's recommendation that "the single most important objective of the condor research team should be to place radio transmitters on every individual in the condor population to the extent this is possible . . ." At the other extreme in a spectrum of opinions about telemetry, the State's advisory committee stated in 1981 that, "Except in the rarest of circumstances, radio tagging itself cannot directly benefit the condors, . . ." and recommended against the uses of any telemetry during the first year of study, with no recommendations on this topic for future years.

With this background of review and disagreement over the condor program, an overall request by the Patuxent Wildlife Research Center, U.S. Fish and Wildlife Service, in March, 1981, for a renewed trapping permit, was not approved by the California Fish and Game Commission until August, 1981. And it was not until January, 1982, that the California Department of Fish and Game reapproved the use of patagial-mounted radio transmitters, and the trapping program could begin. The renewed permits allow for an initial trapping period lasting from 15 September 1981 to 31 May 1982, with only subadult or immature-plumaged condors to be trapped after January, 1982. During this trapping period, three immature or proven, non-mated adult condors may be trapped for the captive breeding program, to include a female as a potential mate for the one captive male California condor at the Los Angeles Zoo, and a male and female for the San Diego Wild Animal Park breeding facility. In addition, the permits allowed for the capture of two additional condors for the telemetry program. The condor research team has the option to reappear before the California Fish and Game Commission to seek approval for an additional four condors for telemetry, if all goes well with the first two birds fitted with radios. Since the permit was not actually issued until January, 1982, it was not possible to trap an adult condor in the short time remaining for trapping adult birds.

Field Studies

In the absence of radioed California condors, the principal efforts in the field during 1980 and 1981 were directed to three objectives. These have been: (1) to locate active nests and to closely monitor the behavior of adult condors and their chicks at these sites, (2) to search for condors in all suitable habitats, evaluating the quality and quantity of data on wild condors that can be collected through the more traditional, naturalistic, field-observation techniques, and determining the usefulness of data collected by these means for answering key questions about the bird, and (3) to prepare for the telemetry program by continuing the field testing the various trapping and radio-tracking techniques and equipment that have been proposed for use on condors. Some results from our attention to objectives 2 and 3 have been reported earlier in this paper.

The search for active nests, and subsequent study of condor behavior at nests that were located, has been the most rewarding and significant of our field activities. These nesting studies were endeavors where we could begin to evaluate suspected condor problems, without having marked birds. It has been suggested that California condors have experienced reduced nesting success during the 1970s, a feeling that was somewhat heightened by the reports of no known nesting attempts in either 1978 or 1979 (Wilbur 1978, and subsequent pers. comm.). If mated pairs could be located early in each nesting season, then close observations of these pairs in the vicinity of their nest caves might reveal the nature of reproductive problems, should such occur.

Four reproductively active pairs of condors were located during 1980 and again in 1981. In 1980, the four pairs consisted of one pair that was occasionally feeding a still-dependent chick that had fledged from a previously undetected nesting in 1979, two pairs that each produced an egg during the normal February-March laying period in 1980, and a fourth pair that we think may have laid and failed. One of the two pairs known to have laid successfully fledged a chick during November, 1980, while the other certain nesting pair, in Santa Barbara County, lost its chick during the July nest visit by the condor research staff. Additional details of the loss of this bird are presented below.

The 1981 pairs consisted of the one pair that fledged a chick in 1980, and which continued to feed that dependent chick into 1981, the Santa Barbara County pair which re-nested in 1981, and a pair in the Sespe Sanctuary that renested after apparently failing in 1980. In addition, we located an apparent fledgling condor during the summer of 1981 from still another, previously unknown 1980 nesting. The Santa Barbara pair apparently, but not certainly, made two nesting attempts in 1981 in caves about 10 km apart. The first was in February, and terminated early during the incubation period, presumably due to destruction of the egg. This site was visited by our staff during the fall of 1981, and found to contain condor egg shell fragments of unknown age. The second nesting attempt was initiated in late April and also ended in failure when the pair lost its egg or newly hatched chick (we don't know which) for an unknown reason at about the time when hatching was due to occur. The Sespe pair successfully fledged a young bird during September, 1981. That chick continued to survive through the end of the year.

Relatively little was learned about the one pair of condors that was feeding a fledged chick early in 1980, because the three birds were so highly mobile, and it was not possible for our field team to monitor most feedings or to track the daily movements of this family group. Nothing was seen in the behavior of these three birds that appeared abnormal, or suggested that there had been any problems in the reproductive effort of this pair.

By far the most valuable observations came from the successful 1980 and 1981 nest sites. We were fortunate in locating the first pair during early March, 1980, and had an observation blind in place and manned on a daily basis beginning a full week prior to egg laying. A daily record of the behavior of this pair and its chick was made from the second week in March until late November, after the young condor had flown from the nest cave.

Some of the key events in the nesting schedule of the 1980 pair are as follows: The pair laid the egg on or about 15 or 16 March, and the egg completed hatching early on the 14th of May. The chick was brooded during all or portions of the day until 13 June, while night-time brooding continued until the end of June. The chick first stood fully upright on its legs on 19 June, first walked out of the nest cave on 2 July, and made its first flight on 7 November.

The data collected from the 1980 nest, along with more limited data from the successful 1981 nesting, specifically on the role of the two adults during incubation and while feeding the chicks, can be compared with similar data collected by Carl Koford in 1939. The degree of difference or similarity between the behavior and daily attentiveness patterns by the adults in the two studies could reveal the nature of reproductive problems, should such have developed since the time of Koford's study. The comparison of attentiveness during incubation by the adult condors at the 1939, and 1980 nests is shown in Table 1, while the comparison for feeding rates for 1939, 1980 and 1981 is shown in Table 2.

These comparisons suggest that the 1980 and 1981 pairs had no more

Table 1. Attentiveness during incubation by nesting pairs of California condors

Variable	Location (year)	
	S–151 (1939)	S–353 (1980)
% time adult "A" in nest cave	73[a]	50[b]
% time adult "B" in nest cave	29[a]	50[b]
Mean no. of exchanges (range) between adults per day, N = days of observations[c]	0.43 (0–2) N = 13	0.23 (0–2) N = 48
Mean length of attentive periods in hours (range) N = no. of observations	31.5 (6–48)[d] N = 4	104.0 (21–216) N = 6

[a]Recalculated from Koford (1953) Table 2 in appendix.
[b]Based upon 624 hours of observations.
[c]For days with continuous observations of 8 hours or more.
[d]Based upon Koford (1953), Figure 11.

problems finding food than did the 1939 pair. The longer intervals of time between nest relief during incubation by the adults in the 1980 nest, by comparison with 1939, at first suggested that each of the 1980 adults was spending relatively more time searching for food than was needed in 1939. But if this was the case, which now seems unlikely, as an explanation for the differences in frequency of adult exchanges, the 1980 pair had no apparent problem finding food once their chick hatched. Both 1980 adults brought food to the young condor on almost a daily

Table 2. July through August feeding schedules for 3 successful California condor nests

Variable	Location (year)		
	S–151 (1939)	S–353 (1980)	S–354 (1981)
Total days of observation[a]	10	61	30
Days with one or more feedings (%)	8 (80%)	44 (72%)	21 (70%)
Mean no. of feedings per day (range)	1.10 (0–2)	1.2 (0–4)	1.0 (0–2)
No. of days adult "A" visited nest	7	29	12
No. of days adult "B" visited nest	7	29	14
No. of feedings by adult "A"	6	39	14
No. of feedings by adult "B"	5	36	16
Mean minutes adult "A" spent per visit (range)	3.2 (2–5)	11.2 (1–33)[b]	12.2 (2–62)[c]
Mean minutes adult "B" spent per visit (range)	6.4 (1–18)	17.7 (3–48)[d]	20.1 (3–65)[e]

[a]Days with 8 or more hours of continuous observation.
[b]Based on 25 feedings, omits upper extreme.
[c]Based on 14 feedings.
[d]Based on 27 feedings, omits upper extreme.
[e]Based on 14 feedings, omits upper extreme.

basis during the early nesting period at a rate similar to the pairs in 1935 and 1981. Complete attentiveness data were not collected from the successful 1981 nest, which was not located until the chick was approximately one month old. The comparison between the 1939, 1980 and 1981 nests reveals no suggestion of behavior change or food shortages. Aside from the obvious problem of inadequate sample size for reaching meaningful conclusions at this point in the study, 2 other factors weaken the value of this comparison. First, Koford watched the 1939 nest at irregular intervals totalling 107 full or partial days between early incubation and post-fledging (March 23, 1939 to March 24, 1940) rather than on a daily basis as was done in 1980 (March 7 to November 26, 1980) and in 1981 (June 16 to October 18, 1981).

Secondly, Koford may have placed his observation blind so close to the nest he was watching that his presence influenced the duration of adult visits into the nest cave (*see* Koford 1953, Table 2). Koford's field notes for 1939 reveal that his blind was 100 to 150 m from the cave entrance (compared to an estimated 800 to 1,200 m in 1980 and 1981), and also that he was not well hidden in the blind. His notes include several references to possible disturbance to the birds due to his close presence (July 3, August 1, and August 13, 1939), including such statements as on August 1, when he wrote, "The very long 3+ hour wait for this bird before entering the nest was probably due to my presence . . ." Thus, the data from the 1980 and 1981 nests may become most useful as a baseline for future nesting studies, rather than for comparison with the perhaps human-influenced observations by Koford in 1939.

The most serious setback to the condor recovery effort was the death of a nestling condor during the summer of 1980, as the bird was being examined by a condor research biologist. The loss of this young bird, one of two condors known to have been hatched that year, was tragic not only because the wild population can ill afford the loss of even a single bird, but also because almost all recovery efforts for the condor came to a halt for 18 months while the State of California, its advisory committees and others redebated the advisability and priorities of the recovery program. A consequence was the State's eventual decision to slow the pace of the program to a much greater extent than was thought advisable by the Fish and Wildlife Service, the National Audubon Society and the major professional ornithological societies, by sharply cutting back on the number of condors to be trapped for telemetry.

The visit to active condor nests during 1980 was intended to be the first in a series of annual studies of growth and development of nestling condors to evaluate the question of whether pre-flight condors are suffering because of inadequate quality and quantity of food. Although the sample size in such studies will never be large enough for statistical treatment of these data, we are convinced that considerable insight into the food resources questions can come from periodic examination of known-aged condor chicks. Looking at the range of growth patterns, even in a small number of chicks, through a combined program of measuring rates of bone and feather development, examinations for signs of malnutrition as evidenced by bone or feather abnormalities, noting the relatively vigor and physical appearance of young birds, could reveal indications of food problems if they exist, within a relatively few years of study. The growth and development study was considered an important segment of a broadly-based study to take a fresh and thorough look at all aspects of condor biology, as a way of evaluating all possible factors that may be contributing to the species' decline.

The autopsy report prepared by the San Diego Zoo hospital (Marilyn P. Anderson, pers. comm., July 1, 1980) showed that the nestling condor's death was due to "Shock, with the cause of death being hypoxia from massive pulmonary edema, secondary to acute cardiac failure. Relative obesity may have been contributory." Later analyses by the Patuxent Research Center showed that tissues from the dead condor were remarkably clean of heavy metal and organochlorine contaminent residues. The best explanation for why the nestling condor died is that it became excessively stressed by prolonged handling, which led to cardiac failure.

Although early indications, based upon the nests watched in 1980 and 1981, are that the pairs of condors watched so far are having no problems with nesting or rearing young, the results of the first two years of intensified field study have yet to produce good evidence of what factor(s) is (are) causing their decline. It remains unlikely that new light will be shed on this question, or that the precise feeding areas being utilized by condors on the unprotected and partially inaccessible (for the condor research staff) private lands, will be adequately identified until a minimum of one-fourth to one-half of the wild population of condors are carrying radio transmitters. Until such time as the recovery program becomes guided primarily by biological design and priorities rather than political considerations, then the achievement of this goal likely will be delayed.

LITERATURE CITED

Carrier, W. D. 1971. Habitat management plans for the California Condor. U.S. Forest Service. 51 pp.

Clark, D. B. (Chairman). 1979. Report of the Sierra Club California Condor Advisory Committee. 5 pp.

Eno, A., J. Spinks and G. Smart. 1979. Recommendations for implementing the California Condor Contingency Plan. U.S. Fish and Wildlife Service Internal Report. 29 pp.

Holden, C. 1980. Condor Flap in California. Science 209:670–672.

Kahn, R. 1980. Death of a condor. Audubon 82(5):14, 16, 18.

Kiff, L. F., D. B. Peakall and S. R. Wilbur. 1979. Recent changes in California Condor eggshells. Condor 81:166–172.

Koford, C. B. 1953. The California Condor. Natl. Audubon Soc. Res. Rep. 4. 154 pp.

Miller, A. H., I. McMillan and E. McMillan. 1965. The current status and welfare of the California Condor. Natl. Audubon Soc. Res. Rep. 6. 61 pp.

Phillips, D. and H. Nash (eds.). 1981. Captive or Forever Free The Condor question. Friends of the Earth, San Francisco. 297 pp.

Pitelka, F. A. 1981. The Condor case: an uphill struggle in a downhill crush. Point Reyes Bird Observatory, Newsletter 53:4–5.

Ricklefs, R. E. (ed.). 1978. Report of the Advisory Panel on the California Condor. Audubon Conservation Report 6. National Audubon Society. New York. 27 pp.

Stallcup, R. 1981. Farwell, Skymaster. Point Reyes Bird Observatory, Newsletter 53:10.

Wilbur, S. R. 1974. Future of the condor: recovery of extinction. Field Mus. Nat. Hist. Bull. 45(9):3–6.

Wilbur, S. R. 1978. The California Condor, 1966–76: A look at its past and future. North American Fauna, No. 72. U.S. Fish and Wildlife Service. Washington, D.C. 136 pp.

Wilbur, S. R. 1980. Estimating the size and trend of the California Condor population 1965–1978. California Fish and Game 66:40–48.

Wilbur, S. R., E. Esplin, R. D. Mallette, J. C. Borneman, and W. H. Radtkey. 1977. A contingency plan for preserving the California Condor. U.S. Fish and Wildlife Service.

Wilbur, S. R., D. Esplin, R. D. Mallette, J. C. Borneman, and W. H. Radtkey. 1980. California Condor recovery plan. U.S. Fish and Wildlife Service. 56 pp.

Verner, J. 1978. California Condor: status of the recovery effort. U.S. Forest Service General Technical Report PSW-28.

Bird Conservation News and Updates

IN this section, the editorial board has selected a few of the noteworthy recent events in bird conservation. Topics vary considerably in scope and include such items as: conservation activities for selected taxonomic groups, legislation that affects bird conservation, habitat preservation that importantly affects bird populations. In each case a member of the editorial board or an expert on the specific topic has summarized the available information.

Convention on International Trade in Endangered Species

The third conference of the parties to the Convention on International Trade in Endangered Species of Wild Fauna and Flora (CITES) took place in New Delhi, India, from 25 February to 8 March 1981. Delegations from 50 out of 65 party nations and 16 nonparty nations attended, as did several international nongovernmental organizations and a number of national, nongovernmental organizations.

Shortly before the conference the Reagan administration changed the membership of the U.S. delegation to reflect more accurately its political philosophies. Four members were replaced, including most of those on the delegation with field experience with the species being protected by the convention, in favor of representatives of state game management agencies and law enforcement officials.

The parties approved several technical improvements to the convention. These included: (1) uniform import and export forms for trade between parties as well as standardized permit requirements for trade with nonparty nations, (2) guidelines for transport of live animals and plants, (3) security paper and validation stamps to reduce permit forgeries, (4) standardized annual report forms to facilitate trade-records analysis, (5) a 10-year review of species listings, and (6) a procedure for disposal of confiscated specimens or products.

Progress was made on the preparation of identification manuals for port inspectors. Several parties are preparing manuals on groups of birds, including ones on the cranes and pigeons by Switzerland, pheasants by the German Democratic Republic, and owls by Denmark. The U.S. offered to prepare one on Caribbean *Amazon* parrots, and the United Kingdom offered one on parrots in general. Five years ago the U.S. government began preparation of a manual on New World parrot identification for port inspectors, but the project was dropped for lack of funding and interest. RARE, Inc. also began preparation of a manual on New World parrots in 1980, but this project was also shelved when hoped-for funds from the U.S. government were not forthcoming.

The U.S. and Canada put forward a proposal to restrict listings on Appendix 2(2)b (organisms controlled because they are not readily distinguished from threatened organisms on Appendix 1 or 2(2)a) to those that can be shown to enter trade. This proposal was withdrawn upon expression of considerable opposition from other delegations.

Support from several countries, but not the United States, was shown for Australia's proposal to consider a "reverse listing" concept for the appendices; that is, listing only those species in which trade is fully justified, rather than those in which trade must be restricted. Impetus for this proposal came, in part, from a resolution of the 1978 ICBP World Conference. A document will be prepared by Australia, the U.S., and the United Kingdom, with requested advice from Mr. Alistair Gammell, ICBP's representative to the conference, to explore the legal and technical ramifications of reverse listing.

Prior to the conference the U.S. Section of ICBP in conjunction with TRAFFIC, USA, proposed a number of new appendix listings to the U.S. government. These were generally accepted and put forward by the U.S. for consideration of the parties. Four neotropical parrots were proposed for Appendix 1 and 21 for Appendix 2(2)a. Following a series of open hearings the U.S. government adopted a position urged by a number of conservation organizations to put forward the listing of all other parrots, excepting the budgie (*Melopsittaca undulata*) and the cockatiel (*Nymphicus hollandicus*), on Appendix 2(2)b. A large number of parrots were already listed prior to the New Delhi conference. The United Kingdom proposed a quite similar listing, adding the rose-ringed parakeet (*Psittacula krameri*) to those excepted from listing. Last minute instructions to the U.S. delegation from the new U.S. administration caused the U.S. delegation to withdraw its proposal and to vote against the U.K. proposal. In spite of opposition from the U.S. and also from Switzerland, the proposal to list all parrots passed easily, with strong support, it should be noted, from third world countries where most of the parrots in trade originate. The four U.S. Appendix 1 parrot proposals—maroon-fronted parrot (*Rhynchopsitta pachyrhyncha*), red-necked parrot (*Amazona arausiaca*), red-tailed parrot (*A. brasiliensis*), and yellow-shouldered parrot (*A. barbadensis*)—were adopted.

A U.S. proposal, also proposed initially by the U.S. Section of ICBP, to list the Mauritius pink pigeon (*Nesoenas mayeri*), among the world's rarest birds, on Appendix 1 was withdrawn by the U.S. Other species added to Appendix 1 included the Humboldt penguin (*Spheniscus humboldti*), white-winged guan (*Penelope albipennis*), and Coxon's double-eyed fig parrot (*Cyclopsitta diophthalma coxoni*). A U.S. proposal to downlist the North American population of the gyrfalcon (*Falco rusticolus*) was grudgingly approved by the parties upon assurance by the U.S. and Canada that the Greenland population was understood to remain on Appendix 1, and that all gyrfalcons entering trade would be marked with a tamperproof band to permit determination of their North American origin.

The U.S. Department of Interior, prompted by strong pressure from the International Association of Fish and Wildlife Agencies and pet industry organizations favored a U.S. reservation on the parrot Appendix 2 listing, but opposing views from the Departments of State and Justice and conservation organizations prevented the reservation from being taken. A U.S. reservation would not have affected requirements on parrots entering or leaving the U.S., and would have reaffirmed a precedent the U.S. had strongly opposed since the convention came into force. The fact that final U.S. positions on several issues reversed positions initially arrived at through a lengthy, and highly commendable, series of public hearings, gave conservationists cause for worry about the receptivity of the Reagan administration to input from the conservation community in the U.S. To have taken a reservation would have eroded still further waning world confidence in the U.S. government's present commitment to wise use and careful stewardship of the world's natural resources.

The U.S. government recently completed its 1979 annual report of species imported and exported under CITES. The report, obtainable from the Government Printing Office, lists a total of 87 Appendix 1 birds, 1,910 Appendix 2 birds and 3,753 Appendix 3 birds imported alive in 1979. A total of 12 Appendix 1 birds and 82 Appendix 2 birds were exported alive in 1979. Dead specimens or products of one Appendix 1 bird and 44 Appendix 2 birds were imported in 1979. If accurate subspecific determinations had been made, it is likely that 60 of the 87 Appendix 1 imports, 1,300 of the 1910 Appendix 2 imports and all 3,753 Appendix 3 imports would not have been listed. No obvious or serious violations of CITES were apparent, taking the report at face value.

WARREN B. KING

Ramsar Convention

The Ramsar Convention on Wetlands of International Importance Especially as Waterfowl Habitat was negotiated in 1971 and came into effect in 1975. It remains the only international wildlife conservation treaty focused on the protection of habitat. By September 1980, 27 nations had become contracting parties and 3 more had signed subject to ratification. The convention requires each contracting party to designate one or more of its wetlands for the convention list of wetlands of international importance. Implicit in this designation is acceptance of the party's responsibility to formulate and implement land use planning for designated wetlands. Unless they compensate for them by creation of new reserves, parties go against the spirit of the convention in delisting wetlands, and should expect to defend their action at conferences of the parties. In spite of its having been labeled a weak convention in that it provides only moral obligations—there are no penalties for withdrawal or misuse of designated wetlands—the Ramsar Convention has been remarkably successful in promoting wise use of wetlands over a substantial portion of the world, particularly in Europe. By September 1980, 214 wetlands covering nearly six million hectares had been designated for the list.

A serious weakness of the convention has been a lack of representation of New World countries. A resolution of the 1978 ICBP World Conference urged nations not yet party to the Ramsar Convention to become so, with special reference to countries in the Americas, and particularly the United States and Canada. It is most encouraging to report that Canada submitted its papers of adherence to UNESCO, the convention's repository, in 1980, and that Chile has expressed a strong interest in becoming a party as well.

In the United States substantial momentum was built toward accession in 1979 and 1980. The Departments of State and Interior initiated action on groundwork documents preparatory to accession, including a draft Environmental Impact Statement, which had entered its fourth draft but was not yet acceptable to State or Interior.

As an indication of the U.S. government's commitment to Ramsar accession, the U.S. sent a team of three observers to the Conference on the Conservation of Wetlands of International Importance, held in Cagliari, Sardinia, Italy, on 24–29 November 1980. The purpose of the conference was to review the functioning of the convention with an eye to making changes, major or minor, in the convention that would improve its effectiveness. Among the significant topics considered at the conference were these: (1) The specific criteria for identifying wetlands of international importance were expanded to include not only wetlands of importance to waterfowl, but also to plants and other animals, as well as representative or unique wetlands; (2) A "shadow" list identifying all of the world's wetlands of international importance was to be prepared by the International Union for the Conservation of Nature (IUCN) and the International Waterfowl Research Bureau (IWRB), the two organizations most directly involved with the convention, as a first step toward attaining a goal of bringing most or all of these wetlands to the convention list; (3) There was general consensus that only minor revisions to the convention should be undertaken, including new provisions for amending the convention, the establishment of a permanent secretariat and predictable funding for the convention, the addition of other official languages than English (France had not acceeded because French was not an official language), and assurances of future participation of non-governmental organizations as observers. Recommendations were passed on these issues; a future meeting of the parties would be required to implement them.

Within the United States initial opposition to accession has come from several state governments, who fear U.S. accession might deprive state wildlife agencies of some of their traditional managmenet and enforcement rights. The federal government's tentative proposal to list as its first wetland of international importance Aransas National Wildlife Refuge, wintering ground for the whooping crane, as a complement to Canada's listing of Wood Buffalo National Park, the Whooping Crane's breeding ground, received strong opposition from the Texas Department of Parks and Wildlife. The Environmental Impact Statement under preparation addressed this issue and another of concern to state governments, whether changes made to the convention after U.S. accession would be binding to the U.S.

Unfortunately, with the arrival of a new federal administration, the useful momentum generated on accession to the Ramsar Convention has been lost as administrators wait for directions from their new supervisors. As of 1981 there is no work being done on the Ramsar Convention within the U.S. government, and a resumption is not expected for two or three years. WARREN B. KING

Fish and Wildlife Conservation Act

The Fish and Wildlife Conservation Act of 1980 was signed into law September 29, 1980, to be implemented October 1, 1981. The Act provides for grants to the States for two major thrusts—the development of conservation plans covering the full spectrum of fish and wildlife (defined as wild, unconfined vertebrates) and the actual management of nongame fish and wildlife (defined as fish and wildlife excluding those normally taken for sport, food, fur or commerce; those listed as endangered or threatened under the Endangered Species Act; those classified as marine mammals under the Marine Mammal Protection Act; and those reverted from a domesticated to a feral existence).

Although authorized at $5 million per year through 1985, it is very doubtful that full funding will be appropriated the first year. The source and level of funds beyond 1985 is subject to further congressional action aided by recommendations from a mandated study to be conducted by the Fish and Wildlife Service.

WARREN B. KING

Federal Endangered Species Program

With no bird species being added or deleted from the U.S. List of Endangered and Threatened Wildlife the totals for birds remained as follows on May 1, 1981: 210 endangered (52 in the U.S.) and 3 threatened (all in the U.S.).

The following recovery plans have been approved by the U.S. Fish and Wildlife Service for birds: eastern brown pelican, California condor (rev.), peregrine falcon (eastern), peregrine falcon (Rocky Mtn.-southwest), Aleutian canada goose, Hawaiian waterbirds (i.e. duck, coot, gallinule, stilt), masked bobwhite, light-footed clapper rail, Mississippi sandhill crane (rev.), whooping crane, California least tern, red-cockaded woodpecker, palila, Kirtland's warbler, and dusky seaside sparrow.

Final drafts are awaiting approval from the Director, U.S. Fish and Wildlife Service for the following recovery plans: bald eagle (southwest), peregrine falcon (Alaska), peregrine falcon (Pacific), yuma clapper rail, and Hawaiian forest birds (rev.).

Semifinal or technical review drafts for the following birds' recovery plans are under consideration: bald eagle (Chesapeake Bay), everglade kite, Puerto Rican plain pigeon, and Puerto Rican parrot.

The I.C.B.P.-U.S. Section submitted a petition to the U.S. Fish and Wildlife Service to list 77 avian taxa. Nineteen were native to the United States (including various territories and the Pacific Trust islands) and 58 taxa were not native to the U.S. This petition was acknowledged and accepted in the *Federal Register* of May 12, 1981, with the petition having been submitted on November 28, 1980.

WARREN B. KING

Restoration of Bald Eagles in New York and Elsewhere

One of the most significant events in the management of endangered species occurred in New York State, when two, wing-tagged bald eagles that had been released by hacking in 1976 paired and nested near Watertown, New York, in 1980. This nesting marks the first time eagles have successfully reproduced in New York since 1973, the last year in which the only other surviving pair had been able to rear a chick.

These successful parents are birds that had been hacked at the Montezuma National Wildlife Refuge at the north end of Cayuga Lake four years earlier, in a pioneering project sponsored by the New York Department of Environmental Conservation and the U.S. Fish and Wildlife Service. Largely through the interest of former Assistant Secretary of the Interior, Nathaniel Reed, New York's Commissioner of Environmental Conservation, Ogden Reid, and Elvis J. Stahr of the National Audubon Society, The Peregrine Fund, Inc. at Cornell University's Laboratory of Ornithology received a contract to help New York State begin its endangered species program by developing a hacking project for young bald eagles.

After obtaining two "runt" eaglets from nests containing three young in Wisconsin, Tina Milburn, a Cornell graduate student, conducted the field work for the first two years in the Montezuma refuge. Both of the 1976 eaglets carried distinctive yellow wing-tags for easy identification in the field. After initial dispersal in the fall, the male returned annually to Montezuma, appearing for a few weeks at a time from late August to October, but no one reported seeing the female again until the fall of 1979, when state biologists found both eagles together at a small lake near Watertown, 80 miles north of Montezuma.

Finding both eagles together after more than three years and nearly in full adult plumage was remarkable, but it soon became obvious to state biologists on the scene that the eagles were actually paired and on territory in an area where eagles had never been known to nest before. In spring the birds built a stick nest in a large red maple, laid eggs, and hatched two young. Later biologists found one of the nearly fledged young dead at the foot of the nest tree. No cause of death could be determined, but it probably fell out of the nest and once below the dense canopy became lost to the parents. The other young flew from the nest and remained in the area with its parents until well into October, the time of dispersal. Peter Nye of the New York State endangered species unit reported that the pair returned to their nest in 1981.

In 1977, Tina Milburn hacked five eaglets (two captive produced at the U.S. Fish and Wildlife Service's Patuxent Wildlife Research Center and three wild-taken) from the tower in the Montezuma refuge. One was shot later in western Pennsylvania, but sightings of marked eagles indicate that at least three of the five 1977 birds were still alive in the fall of 1980.

Meanwhile, endangered species biologists working under the supervision of Peter Nye were busy with the only surviving "wild" pair of bald eagles remaining in the state at a long occupied but seldom successful nest south of Rochester. This pair had last produced a chick in 1973 and had raised only three in the previous ten years. Addled eggs from this nest have been analyzed for chemical residues and are among the most contaminated raptor eggs ever reported, especially for DDE. Recently, however, this old, reproductively malfunctional pair had served as successful foster parents of captive produced eaglets placed in their nest in 1978 and 1980.

In January of 1981 the body of an adult, male bald eagle was found about a mile from the nest. The bird had died of infection resulting from wounds inflicted by two shotgun pellets. When the surviving female had not returned to the nest by late February, the usual time for the eagles to arrive on their territory, biologists began to fear that the history of this beleaguered nest had come to an end. Three weeks later she did return with a new male, and his wing-markers revealed that he was one of the birds hacked in 1977 at the Montezuma refuge. On 8 April, this new pair, after incubating an artificial egg (the female has not laid an egg in the last three years), was given two 21-22-day-old, Patuxent-reared chicks to fledge, and the adoption was an immediate success.

The rate at which hacked eagles have become established as breeders in the wild is well nigh unbelievable. Of the seven eagles hacked by Tina Milburn and The Peregrine Fund, Inc. in 1976 and 1977, three are now breeders, and at least one other is probably still alive after four years. These figures augur well for the establishment of additional pairs from the total of 20 eaglets released at Montezuma under the State's endangered species program through 1980.

The reintroduction to the wild of captive-bred or translocated band eagle nestlings by hacking has suddenly become a popular management tool. New York plans an ambitious expansion of its program in "Phase II"—to release 129 eaglets over the next five years, 20 to 30 a year. If done carefully this project might result in the establishment of 20 or more pairs in New York State. In addition, several other states or regional agencies have started or plan to start programs patterned after New York's operation. The TVA hacked two eaglets from a tower in the Land-Between-the-Lakes area of western Tennessee. Six nestlings were translocated from the San Jaun Islands of Washington to repopulate Catalina Island in southern California in a project jointly funded by the Catalina Island Conservancy, California Department of Fish and Game, and the Western Foundation of Vertebrate Zoology.

Where are all these eaglets for hacking going to come from? Not from captive breeding programs, for there are too few eagles being bred in captivity, and propagating bald eagles is even more expensive than propagating peregrine falcons. They can most easily and economically come from wild nests. The greatest concentrations of highly productive wild nests are in Canada and Alaska.

In 1976 when we first began looking around for a source of eaglets for the New York State project, I naively suggested Alaska, where thousands of bald eagles nest every year, and where the taking of several hundred young a year would have absolutely no impact on the breeding population. This idea was immediately rejected by the armchair biologists in Washington, D.C. who are enamoured by the pseudo-philosophy about "genetic purity of the races." It was feared that the big bald eagles of Alaska would be genetically so discordant with eagles native to the northeastern USA that transplants from Alaska, if they became established, "could affect the fitness and evolutionary direction of already existing populations" although no one has ever demonstrated that measurable genetic differences exist, and the size difference may simply be a nutritional effect of a richer food supply for eagles in Alaska and Canada. Other potential problems were alluded to: possibly spread of pathogens and parasites, the only really serious possibility; physiological and anatomical changes; behavioral incompatibilities between transplants and natives; and so on. In short, all the old arguments that had to be surmounted in order to release non-indigenous peregrines in the eastern USA surfaced again.

Recently, in a remarkably balanced and well-reasoned "Intra-Service Section 7 Consultation, Federal Aid, Endangered Species Grant, NYE-1-5" signed by Harold J. O'Connor, Acting Associate Director, to the Chief, Division of Federal Aid, FWS dated 18 February 1981, the Office of Endangered Species has stated that "we cannot conclude that transplanting individuals (including Alaskan birds) from one population into the former range of another would pose a threat to the species as now listed. Therefore, it is my biological opinion that your action, as proposed, is not likely to jeopardize the continued existence of the bald eagle and, if successful, may promote the conservation of this species." Natural selection, out-breeding, and heterosis have prevailed over genetic purity, inbreeding depression, and rigid adaptation to local environments! Legal access to Alaskan eagles is now a reality, and New York has made arrangements to obtain the first nestlings in 1981.

Tom J. Cade

Snake River Birds of Prey Area

By now almost everyone knows that an 80-mile stretch of the Snake River Canyon south of Boise, Idaho, provides eyrie-sites for the densest nesting population of eagles, hawks, falcons, and owls known anywhere in the world. On average more than 500 pairs of 15 species of raptors return to the canyon each year to nest and rear their young, including some 50 pairs of golden eagles and 300 pairs of prairie falcons. In 1971, Secretary of the Interior, Rogers C. B. Morton, working in concert with then Governor of Idaho, Cecil D. Andrus, set aside by executive order 26,000 acres of the Snake River Canyon as a protected nesting area for raptors under the jurisdiction of the Bureau of Land Management. It became known as the Snake River Birds of Prey Natural Area.

Writing as Secretary of Interior in June of 1979, Cecil Andrus said, "We were well aware then that the 26,000 acres set aside would protect only a portion of the nesting area—a small 'bedroom' for some of the many birds of prey along the Snake River. To assure survival, the birds would need not only additional 'bedroom' space, but an adequate 'dining' area as well. It is not enough to protect their nests we must also guarantee that they have enough protected hunting area to provide the prey they require for survival."

BLM and a number of contractors set to work to learn how much area the birds of prey nesting in the canyon use as a foraging grounds. It soon became evident that Townsend's ground squirrels and jackrabbits are the principal prey species of the raptors and that these mammals thrive mainly on the grazed sagebrush-grasslands to the north of the river. Irrigated croplands, by contrast, held little in the way of suitable prey. Some of the raptors, particularly prairie falcons, forage as far as 15 miles from their nests.

By 1979 BLM was able to issue a detailed report on the environmental requirements of the nesting raptor assemblage in the Snake River Canyon. The studies indicate that to maintain stable populations through time the raptors need some 600,000 acres of hunting range. The Bureau recommended that Congress be requested to amend Title VI of the Federal Land Policy and Management Act of 1976 to establish the Snake River Birds of Prey National Conservation Area, which would encompass about 720,000 acres, 515,000 acres of public lands, some state land, and some private inholdings.

In May of 1980 Interior Secretary Andrus made the formal proposal that Congress establish this area to protect 609,582 acres of canyonlands and adjacent sagebrush desert. The proposal included 490,685 acres of public lands, 40,428 acres of State land, and 78,469 acres of private lands. Basically, the only human land use that would be curtailed in the Area would be the further conversion of public rangelands to privately owned, irrigated farmlands through the Desert Land Act and the Carey Act, which allow a man and his wife to claim up to 640 acres of free public land if they can show that they can irrigate it. Livestock grazing would continue, military uses would continue, hard rock mineral, oil, gas, and geothermal leasing would all still be allowed. Senator Henry M. Jackson and Congressman Morris K. Udall introduced companion bills to bring the legislation before the Congress.

Partly owing to misunderstandings, partly to misrepresentations by vested interests, and partly to the popular reaction against federal authority known as "the sagebrush rebellion," considerable regional opposition mounted against the proposal, and Senator James McClure, who is now chairman of the strategic Committee on Energy and Natural Resources of the U.S. Senate, became a leading opponent. The original bills did include some 100,000 acres of desert entry lands that had already been filed on but not yet proved, and these lands, already largely ploughed up and destroyed as natural habitat, became a particular focus of controversy. The bills died in committee in the 96th Congress.

As a last resort before the change in administrations, Andrus did the only thing left under the authority given to the Secretary of Interior by the Federal Lands Policy and Management Act: On 21 November 1980 he withdrew 482,000 acres of public land to be set aside for the protection of birds of prey and their habitat until the Congress acts in the public interest. This order effectively withdrew 64,865 acres from the operation of mining laws and 417,775 acres from the disposal of the agricultural land laws for a period of 20 years, or until revoked by another Secretary of the Interior. The Secretary's order could have terminated after 90 days if the House and Senate had passed a joint resolution in opposition; but that did not happen, and the fate of the birds of prey area for the moment lies in the uncertain hands of the new Secretary, James Watt, who is no friend of nature conservation and who is adamantly opposed to further expansion of federally protected lands. He could rescind the order at any time.

The current strategy of conservationists is to keep public interest in this area at a high enough pitch so that an order of revocation would be politically damaging to the new administration, while at the same time renewing efforts to get the legislation through Congress, no easy task with the prevailing political conservatism sweeping through the country like a dry wind from the west. In the long run, the birds or prey nesting in the Snake River Canyon may be saved by the unfavorable economics of pumping water up from the river to irrigate cleared sagebrush land.

TOM J. CADE

Harris' Hawks Returned to Southern California

The Harris' hawk (*Parabuteo unicinctus*) had been extirpated as a breeding species in California by the early 1950s. It formerly nested in limited numbers along the lower Colorado River and in the Imperial Valley, but habitat loss from damming the Colorado and the cutting of riparian woods eliminated the last nesting sites. Now suitable habitat is being restored under coordination by the Bureau of Land Management.

In 1979 the Santa Cruz Predatory Bird Research Group began releasing eyas, juvenile, and adult Harris' hawks in areas where the species once bred. By April of 1981, forty-five hawks had been released by hacking or cross-fostering in the nests of red-tailed hawks. Released birds have come from two sources: Eyasses from captive breeding pairs held under permit by California falconers and, also, juvenile or adult birds that falconers were no longer using for hunting. One pair of adult hawks released early in 1980 set up a territory, built a nest, laid eggs, and successfully fledged two young. In the spring of 1981, at least two pairs are on territory along the Colorado River.

Four captive produced prairie falcons have also mated with wild birds at eyries in California.

TOM J. CADE

Peregrine Recovery Efforts

Successful efforts to restore peregrine falcons to former breeding haunts continued on several fronts in North America during the past year. Since the eastern U.S. recovery program is covered in detail elsewhere in this volume, there is no need for further mention of it here.

In the Rocky Mountain states, William Burnham and his crew at Fort Collins, Colorado, made significant strides in implementing various aspects of the recovery plan for that region, with help from the states of Colorado, Utah, Wyoming, South Dakota, and Nebraska, and from federal agencies, the Fish and Wildlife Service, Bureau of Land Management, Forest Service, and National Park Service. They released a total of 58 young peregrines in 1980. Sixteen young were placed with wild breeding adults at four eyries in Colorado; some were cross-fostered under prairie falcons in South Dakota and Nebraska; and the remainder were hacked at sites in Colorado, Utah, and Wyoming. Six birds had to be recalled owing to problems with local predators, but the field workers felt that 43 of the remainder (83 percent) survived to become independent hunters.

Each year some of the young come from wild eggs brought into the laboratory. Personnel of the Colorado Division of Wildlife removed 19 wild eggs from four active nests in 1980, and of the 13 viable eggs that arrived at Fort Collins, The Peregrine Fund was able to hatch eleven. Eight survived to be released.

Also in 1980 several falcons released in earlier years returned to sites where they could be identified. One 1978 male bred with a wild female at an eyrie in southern Colorado; they raised four young. Several subadult peregrines were seen around other eyries in Colorado, and they were almost certainly released birds too. A male hacked at a cliff in the Rocky Mountain National Park returned for the second year in a row, and early evidence suggests that he is back in the spring of 1981 as well. In Utah three subadults were present around towers where hacking has been underway. Early spring observations in 1981 indicate that released birds are present at hack sites in Rocky Mountain National Park, in northern Utah, and in Jackson Hole, Wyoming.

On the West Coast, the Santa Cruz Predatory Bird Research Group under the direction of Brian Walton has done important work with the surviving peregrines in California. There are about 35 to 40 occupied eyries in California, but many pairs still suffer reproductive failure from DDT-induced egg-shell thinning. To combat this loss, SCPBRG workers collect the thin-shelled, wild eggs and incubate them in the laboratory, as Burnham and his group have been doing in Colorado. Later, the incubator-hatched young are returned to the nests of wild falcons.

This work has been going on for three years in cooperation with the California Department of Fish and Game. Each year the success in hatching the thin, improperly evaporating wild eggs has improved. In 1980 eggs were taken from seven nests where the pairs had failed to hatch any eggs in several years. From 20 hatchable eggs, fifteen young were produced; nine were returned to nests, and six were kept for future captive breeding stock.

By the end of April 1981, Walton was able to report the existence of two new pairs of peregrines on the California coast, both consisting of birds that had been fostered in previous years. This brings the number of occupied coastal sites to six. Fifteen young (captive produced or hatched from wild eggs in the lab) had been fostered into seven wild nests, and Walton expected to be able to place about 15 more before the end of the season. Thirty-two productive pairs were present in the wild in California, and several other sites had unproductive peregrines around them. About 90 percent of these eyries are recently discovered, and very few of the historically known sites in central and southern California have been reoccupied. The two recent coastal sites are notable exceptions.

The Canadians have also achieved encouraging results in their efforts to bolster diminished peregrine populations. The Canadian Wildlife Service project directed by Richard Fyfe produced more than 90 young at Wainwright, Alberta, in 1980. Together with offspring from private breeders, Fyfe and his coworkers released about 100 young to the wild last year (both captive produced young and young hatched from wild eggs). These birds were either hacked or fostered in the nests of wild peregrines. At least seven birds released in prior years were paired and breeding in the wild in northern Alberta in 1980, including one female that had been hacked on a building in Edmonton, Alberta but then dispersed to join the breeding population to the north, where nine pairs are now nesting in an area where only three pairs were present in 1975.

TOM J. CADE

Status of Whooping Crane Conservation Efforts

Whooping cranes (*Grus americana*) exist in three populations: the relict Canadian population that breeds in Wood Buffalo National Park and winters in coastal Texas, the introduced Rocky Mountain population established by placing whooping crane eggs in sandhill crane nests in Idaho, and the captive flock predominantly situated at the Patuxent Wildlife Research Center (PWRC), Maryland. The 1981 total count of whooping cranes stands at 114 birds. There are 78 in the Canadian flock, 14 in the Idaho group, and 22 in captivity. The whoopers have made a remarkable recovery from a low of 21 wild birds in 1941.

The Canadian Wildlife Service (CWS) annually monitors whooping cranes in northern Alberta and the southern portions of the Northwest Territories. Since air surveys were expanded in 1967, CWS biologist Ernie Kuyt has observed the number of nesting pairs increase from 9 (1967) to 19 (1980). Seventeen nests were located in 1981, similar to the count in 1979.

Since 1977 most of the chicks in Canada have been color-marked. Helicopter surveys first locate the family groups. Biologists then land near the chicks, capture the birds and mark them with bright, plastic legbands. Nine chicks were banded in 1977, eight in 1978, six in 1979 and six in 1980. When the whooping crane chicks are captured, blood samples are taken and sent to Dr. Brian Biederman at the University of Calgary for sex determination by chromosome morphology. With both the age and sex of marked birds known, field researchers can gain greater insight into the biology of wild whooping cranes.

Two of the pairs which nested in 1980 contained one member color-marked in 1977, indicating that some whoopers may be able to breed when three years old. Data from PWRC suggest that the marked nesting whoopers may have been males, since three-year old captive males have produced semen samples, whereas, females have never laid at such an early age.

Kurt Johnson, of the Department of Wildlife Ecology at the University of Wisconsin-Madison, made an exhaustive study of whooping crane migration through the Dakotas, Kansas, Nebraska, Oklahoma and Texas in an effort to determine critical migration habitats in the USA. His conclusion was that en route through the USA, whoopers pass quickly from areas in southern Canada, where they linger for several weeks, to winter habitats within the Aransas National Wildlife Refuge in Texas. The cranes usually migrate in small groups, family units, or in conjunction with large flocks of sandhill cranes, and stop to rest overnight or for several days in wetlands where shallow water and a reasonably unobstructed view of the surrounding countryside are available. Wide, shallow rivers, alkali lakes, and large natural wetlands constitute the preferred resting areas for whoopers in passage. The cranes will also often use small ponds and surrounding agricultural fields on migration. Cranes do, however, require wetlands as stopover habitat during migration, and additional research should be undertaken to identify specific wetlands that need to be protected within the migration coridor. Mr. Johnson suggests that private conservation groups like The Nature Conservancy or the National Audubon Society protect these wetlands through purchase of easements in view of current federal policy towards habitat conservation.

Canadian and U.S. biologists will be involved in radio tracking whooping crane chicks on migration from Wood Buffalo National Park to the Aransas National Wildlife Refuge in the Autumn of 1981. Small radios will be attached to the plastic leg-bands placed on 1981 chicks. The objective will be to locate and characterize every stop the families make on migration, and also gain additional information on daily flight distances, exact migration routes, dangers on migration, and habitats that should be protected.

Since 1970 David Blankenship, staff biologist of the National Audubon Society, researched the biology of whooping cranes wintering on the Texas coast. Mary Bishop, consultant to the Society, joined the project in 1980. Both researchers are now concentrating on the behavior of the banded cranes. During the winter of 1980–81 an average of 45 birds resided on the Aransas National Wildlife Refuge proper, an average of 13 birds on federal or National Audubon Society lands on Matagorda Island, and an average of 20 birds on private lands not officially protected.

The second group of whooping cranes, the introduced Rocky Mountain population, continue to increase in response to whooper eggs being placed in the nests of great sandhill cranes breeding at Grays Lake National Wildlife Refuge, Idaho. The Canadian whoopers usually lay two-egg clutches but typically rear only one chick. Consequently, removing one egg from each two-egg clutch does not impede the productivity of the nesting pairs. Since 1967, 103 eggs have been collected from the wild whoopers, of which 50 were sent to the PWRC. Twenty adults and offspring currently survive at the PWRC.

The other 53 eggs collected in Canada two were addled and the remaining 51 were placed in the nests of sandhill cranes in Idaho. In addition, 38 eggs from captive whooping cranes at the PWRC have been placed under sandhills in Idaho, for a total of 89 eggs transferred from 1975–80. Fourteen whooping cranes survived in the Rocky Mountain flock through the winter of 1980–81. Major causes of mortality in the Grays Lake population have been unavoidable due to inclement weather conditions at the time of hatching and coyote predation of prefledged chicks.

From 1976–78 serious problems occurred with eggs transferred from captivity to the wild. Many of the eggs were infertile or the embryos and chicks were weak and died. Since 1979, however, researchers at the Patuxent Center have incubated their whooping crane eggs under captive sandhill cranes, rather than in artificial incubators, before transfer to the wild. Six of the seven eggs incubated under the regime and transferred to the wild hatched and three chicks fledged, a survival rate equal to that of eggs collected from wild whoopers. The Patuxent team is to be congratulated for these steady advances in the captive management of cranes. Their efforts have culminated in captive-produced eggs hatching and growing into adult whooping cranes flying free over the Rocky Mountains.

Whoopers reared by sandhills are successfully adapting to the more upland feeding niche of their foster parents. For example, in winter the western group of whoopers feed in dry agricultural fields, in contrast to their conspecifics in the Canadian population that prefer aquatic animal food in brackish ponds along the Gulf Coast. The Rocky Mountain whooper chicks follow their foster parents during their migration over 800 miles from Idaho south to wintering grounds in the Rio Grande Valley of New Mexico, mainly the Bosque del Apache National Wildlife Refuge. In subsequent migrations, the whoopers repeat the same route, indicating that migration is learned in cranes. Although most subadult whoopers scattered over a 50-mile radius from their natal wetlands at Grays Lake, the older cranes tend to return to the refuge as they approached sexual maturity. Between 1978 and 1980 three 1975- and 1976-produced males established territories at Grays Lake. In so doing they excluded several breeding pairs of sandhills from their defended turf.

As in most crane species, the sex of an individual can usually be determined by the trumpeting "unison call" display, typically performed by a mated pair but also given by single birds in threat encounters. A male whooper emits a series of clear, continuous calls while the female simultaneously produces a series of short abrupt calls—two or three such calls produced for each call of the male throughout the duet. The performance of solo "unison call" by 1975 and 1976-produced whoopers in the Rocky Mountain population indicated that they are all males.

On April 25, 1981, a moulting and, thus, flightless three-year old captive (sandhill-reared) female whooping crane was transported from Patuxent to Grays Lake and placed in an aviary within the territory of one of the male whoopers. Approximately 10 days later she was released. However, no strong attachment existed between the male and the female, although the female has adjusted to the wild beautifully and remains in the male's territory. Interestingly, the arrival of the female was correlated with increased territorial behavior on the part of the male—he has increased his territory size nearly two-fold since her arrival, even though his territory size had remained stable the previous three years.

The sandhill-reared whoopers do not appear to be imprinted on sandhill cranes. Although they accompany the sandhill flocks on migration and winter in company with thousands of sandhills, they are dominant over the sandhills and usually ignore them. The whoopers normally remain aloof from each other (perhaps a reflection of the dominance of males), although they did occasionally gather together on the wintering grounds in 1980 and 1981. Interestingly, older whoopers are sometimes attracted to a whooper chick that is in the company with its sandhill foster parents.

The Canadian Wildlife Service, and particularly Ernie Kuyt, is to be congratulated for faithfully, and at great cost, providing whooping crane eggs to both Patuxent and the Grays Lake experiment. Dr. Roderick Drewien of the University of Idaho also deserves merit for managing the Rocky Mountain experiment. Dr. Drewien has been studying sandhill cranes at Grays Lake and elsewhere since 1969 and patiently yet enthusiastically continues to develop the new migrating whooper flock in the west.

Having developed the captive propagation program for whoopers at Patuxent since 1966, Dr. Ray Erickson retired in 1980. Dr. George Gee and Dr. Scott Derickson now manage 19 whoopers held at the center. Six mated pairs were present in the spring of 1980. Six eggs were laid from which one chick was reared by foster parents at Grays Lake. The 1980 breeding season was seriously affected by several unavoidable disturbances during the nesting period, and the pairs are expected to do much better in 1981.

In January, 1980, a Whooping Crane Recovery Plan, drafted by crane biologists and U.S. officials, was approved by the Director of the U.S. Fish and Wildlife Service. This plan calls for: (1) increasing the Wood Buffalo-Aransas population to at least 40 nesting pairs; and (2) establishing two additional self-sustaining populations of at least 20 nesting pairs each. When this is achieved the species' status will be reduced from Endangered to Threatened. To achieve these goals the Recovery Plan outlines a variety of management and research objectives including habitat management, captive propagation, reduction of mortality rates, and improved public awareness.

When the Rocky Mountain whooper population is self-sustaining and has increased to 20 pairs, efforts will perhaps concentrate on establishing both a nonmigratory population in Florida using the Florida sandhill crane as a foster parent, and initiating a new migratory whooper flock in southwestern Ontario, Canada, where a population of greater sandhill cranes are tentatively designed as surrogate parents for whoopers. Since Great Lake sandhills migrate to Florida in winter, the new migratory flock of whoopers in the east would be allopatric from both the traditional flock and the Rocky Mountain flock, although they can be expected to meet with whoopers reared in the sedentary situation in Florida. The Canadians are keenly interested in embarking on the Ontario sandhill project as soon as the Grays Lake whoopers breed. But, captive whoopers at PWRC sometimes lay so early in spring that their eggs hatch before greater sandhills in Idaho or Ontario nest. The Florida sandhills, however, begin nesting in February and March and thus are potential candidates for the early eggs from PWRC.

GEORGE W. ARCHIBALD

Conservation Activities for Parrots

The ICBP World Working Group on Parrots organized a conference and workshop on parrots in Castries, St. Lucia, on 17–19 April 1980. Conservation problems in the new world tropics and in particular the islands of the Caribbean were stressed. The 70 participants included representatives of 16 countries. Working Group members prepared a priority program for parrot conservation in the New World tropics, within the framework of ICBP's World Bird Conservation Priorities (see ICBP Newsletter Vol. 3, No. 1, January 1981). The conference proceedings, including this priority program and the texts of 29 papers, will be published in 1981 by ICBP as the first number in its new Technical Publications series.

The conference venue and program provided an excellent focus on the severe conservation problems confronting the St. Lucia parrot (*Amazona versicolor*) and the other three endangered *Amazon* parrots of the Lesser Antilles. The St. Lucia Forestry Division has made particularly encouraging strides in conserving the St. Lucia parrot and the rainforest on which it depends through an intensive program of conservation education, research and enforcement. Regrettably, a severe hurricane crossed Dominica and St. Lucia several months after the conference on 4 August 1980, damaging up to 80 percent of the forests of both countries and setting back St. Lucia's exemplary conservation efforts by several years. Subsequent surveys have suggested that although some loss of parrots undoubtedly took place, about 125 St. Lucia parrots remained. The effect of the hurricane on the parrots' reproduction has not yet been determined, although it may be assumed that the loss of food and nest sites associated with leaf and fruit stripping and uprooting of trees in the 115 mph winds recorded will depress productivity at least temporarily.

Shortly after Hurricane David and Frederick struck, ICBP sent Noel and Helen Snyder to Dominica to review the extent of damage to that island's forests and its populations of imperial parrots (*A. imperialis*) and red-necked parrots (*A. arausiaca*). The Snyders recommended a program of scientific exchange involving training of Dominican Forestry Division personnel at the U.S. Fish and Wildlife Service Field Station in Luquillo Experimental Forest, Puerto Rico, where Jim and Beth Wiley supervise work on the Puerto Rican parrot (*A. vittata*). With the help of World Wildlife Fund, U.S. Appeal, this exchange was undertaken. Two Dominicans, Quammie Greenaway and Nelson Green, received training in Puerto Rico from April to June 1980, and a third Dominican, Michael Zamore, who is in charge of forest surveys, enforcement and wildlife for the Dominican Forestry Division, was present May-June 1980. The three Dominicans returned to Dominica with urgently needed field gear. In October 1980 Wiley went to Dominica where he helped the Forestry Division set up observation towers, develop census techniques, and make further studies of avifaunal changes following Hurricane Allen. The Dominican Forestry Division has a total of 6 personnel working full or part time on parrots, and had two imperial parrot nests under surveillance for the 1981 breeding season. Further exchanges are contemplated for 1981 but may run into trouble due to U.S. federal budget restrictions preventing foreign travel and to increased activities in Dominican forests of the "Dreads," a local group of violence-oriented dissidents.

Work on the Puerto Rican parrot continues to produce results, but recovery has been slow and the long-term prospects remain very much in doubt. Foremost among concerns is the question of the level of commitment of the agencies sharing primary responsibility for the parrot's conservation, the U.S. Fish and Wildlife Service and the U.S. Forest Service.

The U.S. Fish and Wildlife Service Endangered Wildlife Program, administered out of Patuxent Wildlife Research Center, Laurel, Maryland, has recently assigned the Puerto Rican parrot project number three in priority among its commitments nationwide, calling for a doubling of its Puerto Rican parrot budget in fiscal 1982. This increase would permit the U.S. Fish and Wildlife Service to hire an aviculturist and to take over full responsibility for the aviaries in Luquillo Forest housing a number of Puerto Rican parrots and Hispaniolan parrots (*A. ventralis*), the latter used as stand-ins to test new avicultural techniques. Until now the U.S. Forest Service has run the aviary.

The Forest Service has a position for a biologist to work as a counterpart to the Fish and Wildlife Service's Jim Wiley, a position that has remained unfilled for a year, due to a Civil Service paradox in which the best qualified applicant is not the highest ranked applicant. This has led to a disproportionate and wholly unrealistic share of the work falling on the shoulders of Wiley and a small crew.

The parrots have fared well in 1981. At the onset of the breeding season at least 19 birds comprised the population, up from 18 or fewer at the start of the past 3 breeding seasons. As in 1980, 4 pairs bred, but two nonbreeding pairs defended nest sites and may in the future take up artificial nest sites provided for them. The female of one of the 4 breeding pairs has laid thin-shelled eggs for the last 4 years in spite of a dietary supplement of calcium made available and used by her in the nest hole. The female suffered a wing injury early in the breeding season and her first clutch was lost, but by May she had recovered, probably too late to recycle with a thicker-shelled second clutch. This female raised foster clutches of 1 and 2 captive-reared chicks in 1979 and 1980 respectively. Both broods fledged successfully and the 1980 chicks were observed as recently as February 1981.

Of the other three pairs, one pair was newly established but replaces a pair that has disappeared. All three pairs have three young each, two about to fledge in May, the new pair about a month behind. The several problems that have beset the parrot population in years past, including thrasher predation, capture of young by humans, warble fly maggot infestations, hawk predation, rat predation, nests lost to storms or bee swarms, have not affected parrots this year.

In the aviary 6 female Puerto Rican parrots produced 22 eggs, 7 of which, laid by two females, were fertile. Four of these hatched and 2 have survived and an attempt will be made to foster these into a nest in the wild. Artificial insemination may well improve the rate of fertility next year. It was attempted this year probably too early in the breeding season. A technique modification, tried on a female Hispaniolan parrot, of removing eggs as they were laid rather than upon completion of the clutch, yielded a remarkable 20 eggs, 18 of which were fertile. If applicable to Puerto Rican parrots, this technique holds promise of vastly increasing the output of captive birds.

What is needed at this point is several seasons of favorable reproduction and especially some new recruitment to the breeding population. With only four pairs breeding and one of these laying thin-shelled eggs, there is little cause for optimism, in spite of a slight increase in the population from its all-time low of 14 in 1971–74.

I would like to thank Randy Perry and Jim Wiley of the U.S. Fish and Wildlife Service for providing information for this report.

WARREN B. KING

Status and Conservation of Woodpeckers

There is no new information concerning the highly endangered imperial woodpecker (*Campephilus imperialis*), endemic to Mexico. No verified reports of ivory-billed woodpeckers in the United States have come to my attention. Dr. Storrs Olson of the Smithsonian Institution reports that discussions by Olson and Dr. Eugene Morton with Cuban ornithologists Abelardo Moreno and Orlando Garrido last November disclosed that the tiny ivory-billed population in Oriente Province, eastern Cuba, thought to be holding its own heretofore, may be in great danger. No one has been able to locate the woodpeckers in the region where they were last known to occur. If corroborated that the Cuban population is seriously depleted or non-existant, the species may be added to the list of extinct North American birds in this decade.

The U.S. Fish and Wildlife Service and U.S. Forest Service are cooperating in a census of the endangered red-cockaded woodpecker (*Picoides borealis*). Although the census was to be completed in spring of 1981 the results are as yet unavailable. Local studies continue to show a decline in numbers of the species (e.g., J. Jackson reports that the Savannah River Atomic Energy Plant "colonies" have diminished from 18 to 2 in the past five years). Birds from three "colonies" in an area of Fort Stewart, Georgia, destined to be denuded of trees for military purposes were translocated to St. Catherine's Island, coastal Georgia, through the combined efforts of personnel of the Edward Noble Foundation, the American Museum of Natural History, the U.S. Fish and Wildlife Service, the Georgia Department of Natural Resources, and the University of Georgia. Although four birds were lost within the first few days of the translocation, the other eight have survived beyond 10 weeks and appear to be in good condition. This woodpecker seems never to have occurred on the island. Hence, successful translocation to an area outside the prior range of the species would be especially significant.

LESTER L. SHORT

Tropical Deforestation and North American Migrant Birds

Moist climax forest covers 5 million square km, or 53.5 percent of Central and South America, and comprises nearly half of all the tropical moist forest in the world. In addition, these neotropical forests contain approximately one-tenth of all species (both plant and animal) on earth.

These forests lie mostly, but not exclusively, in developing nations, and there are great pressures to exploit these resources based upon short term benefit analysis. Among the primary factors affecting these forests, conversion by local farmers using slash and burn techniques accounts for 20–50,000 square km annually. Timbering, particularly along waterways, accounted for 20 million cubic meters of hardwood in 1973, and is projected to account for up to 118 million cubic meters by the year 2000. Even when only particular trees are culled from a forest, the actual logging procedure does considerable damage to about one-half of the remaining trees. Clearing for cattle raising accounted for 80,000 square km in Brasil alone between 1966 and 1978. As a consequence of these various factors, these forests are disappearing at a rate of 1–2 percent annually, so that by the year 2000 one-half to two-thirds of existing neotropical forests will be lost, and with them perhaps 3–5 percent of all the world's plant and animal species (Lovejoy 1980).

Of the 650 species composing the avifauna of the contiguous United States, 332 or 51 percent migrate annually to the neotropics. Of these, 107 species live primarily in the neotropical forests during our winter. Approximately 230 species actually migrate through them, spending one-half to two-thirds of their life cycle in the neotropics.

The traditional view has been that migratory, forest passerines use secondary vegetation on their wintering grounds and would not be particularly affected by tropical deforestation, but it has become increasingly apparent that the story is far more complex (Keast and Morton 1980). Indeed, our migratory birds, even the migratory forest passerines, can hardly be treated as a unit, except in the sense that they all migrate. Some species, such as the Kentucky warbler, depend very much on mature tropical forest. Others, such as the Kirtland's warbler, are scrub habitat birds; yet others move from habitat to habitat. It is clear in any case that those which use forest habitat or secondary vegetation will be affected by extensive conversion of the landscape to pasture.

Evidence indicates that even just-logged forests can only sustain 60 percent of their species (Rappole, Morton and Lovejoy 1981). This com-

Table 1. Deforestation rates in several Latin American countries

Country	Total land area (km²)	Present forest cover (km²)	Deforestation rates/year	
			Sommer (1976)	Myers (1980) or others
Mexico	1,963,000	400,000	—	16,000 km²
Honduras	112,000	70,000	—	—
Guatemala	109,000	53,000	—	1,325 km²
El Salvador	21,300	2,600	—	—
Nicaragua	148,000	64,000	—	1–2,000 km²
Costa Rica	49,000	15,000	60,000 km²	300–500 km² but now less
Panama	75,500	30,000	—	—
Colombia	1,138,000	364,000	250,000 km²	22,000 km² (1966–72) 14,600 km² (1972–75) 5,000 km² (1975–76)
Venezuela	916,000	352,000	—	—
Ecuador	300,000	180,000	50,000 km²	—

bined with the high annual rate of deforestation in the neotropics would seem to presage an imminent crisis for our North American migrant birds, particularly for those wintering in Central America and northern South America (see Table 1).

Yet curiously, nationwide breeding bird censuses do not show as many declines as one might expect (Robbins, pers. comm.); indeed censuses of some bird species show increases. Since these censuses only date to 1966, it is possible that populations of birds such as Parulids may have been low at that time, due to pesticide impact on the environment, but are unusually high today because of a major bulge in food availability from spruce budworm outbreaks.

A number of initiatives have been taken on the tropical deforestation issue. The State Department held a meeting in June 1978, ultimately resulting in the production of a report by the Interagency Task Force on Tropical Deforestation, and Presidential notice in the Environmental Message of 1980. The United Nations Environment Program held an experts' meeting on the topic in February 1980. In 1981 World Wildlife Fund-U.S. has declared its intention to develop a program to address this problem, and in the past several months has even added extra staff to work on this issue.

The Western Hemisphere Convention may be a useful instrument to promote hemispheric cooperation in protecting our migrants and the forest vegetation on which they depend, even though the neotropical forest/migratory bird relationship is as yet imprecisely known. Fundamental to that potential is the recognition that, while from a U.S. view they can be thought of as U.S. migrants, this is really a concern about birds of Latin America and the Caribbean which just happen to spend a few months for breeding purposes in North America. They play important roles in the neotropical forest communities, aiding in pollination and dispersal of a number of plant species. These migrants play important ecological roles at both ends of their journeys and can only do so when the full extent of their annual cycles is secure.

THOMAS E. LOVEJOY

REFERENCES

Keast, A. and E. S. Morton (eds.). 1980. Migrant birds in the Neotropics: ecology, behavior, distribution, and conservation. Smithsonian Inst. Press, Washington, D.C. 576 pp.

Lovejoy, T. E. 1980. Global changes in diversity. pp. 327–332. *in* The Global 2000 Report to the President. Council on Environmental Quality and Department of State, Washington, D.C.

Myers, N. 1980. Conversion of Moist Tropical Forests. National Academy of Sciences Washington D.C.

Rappole, J., E. S. Morton and T. E. Lovejoy. 1981. Nearctic Avian Migrants in the Neotropics. Unpublished report to the U.S. Fish and Wildlife Service (Cooperative Agreement no. 14–16–0009–79–942 with World Wildlife Fund-U.S.).

A Status Report on the Dusky Seaside Sparrow

The dusky seaside sparrow (*Ammospiza maritima nigrescens*) inhabited the vast expanse of cordgrass (*Spartina bakeri*) marshes formerly found on the northern part of Merritt Island and still present, although greatly reduced and modified, in the valley of the St. Johns River in Brevard County. Out of a population that once numbered several thousand pairs at both sites, only five aging (5 to 9-years) males survive. As of July, 1982, these birds are being maintained as captive birds. One color-banded male, last seen on 23 July, 1980, has never been seen again.

The Dusky Seaside Sparrow Recovery Team has repeatedly urged the U.S. Fish and Wildlife Service to authorize a captive propagation program to breed these surviving males with female seaside sparrows from the Florida Gulf coast populations of Scott's seaside sparrow (*A. m. penisulae*), or the Wakulla seaside sparrow (*A. m. juncicola*)—both dark-plumaged races that are probably most closely related to the dusky. Indeed, an experimental cross of a Scott's female with one of the dusky males in 1980 produced three young, one of which is a very dusky-like individual.

The recommendation of the Recovery Team, after the initial intra-specific cross, is to backcross the dusky males with the F_1 females (50% dusky) in the second year to increase the dusky percentage to 75%. During the third year the backcross with the 75% birds would increase the dusky line to 87.5%, the fourth backcross-93.8%, and the fifth backcross-96.9%. A phenotypical dusky may be apparent by the second backcross, and may be sufficient for propagation at that point if the original dusky males are no longer available for further backcrossing.

To date, every proposal to conduct this breeding project at no or very little cost to the federal government has been denied by the U.S. Fish and Wildlife Service, using a variety of specious reasons: (1) No assurance that hybridization would produce dusky-like birds, (2) Hybrids may not be fertile, (3) Birds may not accept marsh habitat, (4) Hybridization will result in a permanent dilution of the dusky deme (for a response to this fear of miscegenation of the races, *see* James, 1980), (5) Purposes of the Endangered Species Act cannot be extended to utilize hybridization as a conservation tool, (6) Approval of a breeding project would set a precedent for hybridization, and (if all the preceeding excuses have not discouraged you) (7) the Service does not want to expand any more funds to restore and maintain the dusky seaside sparrow habitat in the St. Johns National Wildlife Refuge (this last statement was made by Acting Director Eugene Hester at the Seaside Sparrow Symposium held in Raleigh, NC, October 1–2, 1981).

We have already lost the breeding seasons of 1981 and 1982 when the captive males came into reproductive readiness, but remain celibate. Instead of beginning the 1983 breeding season with a supply of birds with an 87.5% dusky genotype, we are resubmitting another proposal to begin the initial intraspecific crosses ("hybridization" is not the correct term). If any biological enlightenment or conservation compassion has reached the inner echelons of the U.S. Fish and Wildlife Service in recent days, perhaps this time the project will be approved.

In the meanwhile, the Dusky Seaside Sparrow Recovery Team has been officially disbanded. The original concept of the recovery teams envisioned them as advisory bodies to remain in existence until the endangered species either recovered or went extinct. Not only would the team prepare the recovery plan, but would oversee its implementation. Apparently, many of the recovery teams, which tend to be advocates for the species they represent, have become sources of embarrassment to the present administration, hence the new policy now is to disband the team as soon as a recovery plan has been prepared and approved. This should conveniently result in less attention being brought to the Department of the Interior's reduced efforts in behalf of endangered species.

Two questions asked most often are: (1) How did the dusky get into this position, and (2) What was the single most important cause and who is to blame. The simplest, most generalized answer to both these questions, of course, is humans extirpated the dusky (with the death of the five surviving captive males extinction will replace the term extirpation) by destruction and alteration of the dusky's natural habitat. On Merritt Island the marshes were diked and impounded for the control of mosquito breeding, beginning in 1963, resulting in total elimination of habitat.

In the St. Johns valley the natural cycles of flooding and drying of the marsh were altered by drainage for agricultural practices beginning in the 1920s, but greatly accelerated in the 1960s. Concurrent with this drainage, ranchers burned vast acreages of marsh during the winter dry season to produce spring forage for cattle. The combination of overdrainage and hot fires broke up the extensive marsh system into islands or pockets of suitable dusky habitat interspersed with large areas of unusable habitat, much of this characterized by invasion of shrubby vegetation, especially saltbush (*Baccharis* spp), waxmyrtle (*Myrica cerifera*), and Brazilian pepper (*Schinus terebinthefolius*).

Beginning in the late 1960s and 1970s several catastrophic events befell the dusky's habitat in the St. Johns marsh. First, in 1969, General Development Corporation dug a deep canal through the midst of one of the largest concentrations of duskies. This was followed in 1970-72 by construction of the Beeline Expressway by the State of Florida. This highway crosses the St. Johns river, then splits into one arm heading northeast towards Titusville, and one arm heading southeast towards Cocoa. This intersection was sited in the midst of another concentration of duskies. The fill to construct the road came in part from the nearby marsh, and the site where once up to a half dozen seaside sparrow territories existed is now a 12-acre borrow pit. This road and canal construction disrupted hydrologic conditions and encouraged still more invasion of upland woody shrubs and trees.

In 1969, Brevard County began construction of a road that cut through several miles of dusky habitat. Most of the borrow for the road came from a ditch dug parallel to the roadbed. In 1970 the U.S. Fish and Wildlife Service began land acquisition of 4,000 acres including this road, to establish the St. Johns National Wildlife Refuge for the dusky. Because of concerns expressed by Brevard County Commissioners about possibly causing flooding of upland property east of the refuge, the Service did nothing to stop drainage of refuge lands by this ditch or any of the other former agricultural ditches until the late 1970s.

In 1975, on recommendation of the Recovery Team, the Service began acquiring habitat in the "Y" of the Beeline Expressway. This purchase included the General Development Corporation canal and surrounding lands. Again, to appease the owners of upland property to the east, no attempt was made to plug this canal and stop its affects on refuge lands. To compound matters, the Service suddenly stopped land acquisition in 1978, leaving an important "inholding" in private hands which has the potential to compromise future dusky habitat restoration efforts.

Cattle ranching has been the main agricultural industry in the marsh area of the St. Johns because of high saline content of the soils. To produce forage ranchers burn the marsh each winter during the driest time of the year. Thus, in December and January vast acreages burn for several days at a time. Except in wet winters, little of the *Spartina* marsh on privately owned or leased lands escapes fire. Fire is a necessary component of the dusky's habitat. But natural fires occur mostly in the summer season from lightning strikes when the marsh is wetter and greener, hence summer burns tend to be patchy and less extensive. Because it is not until the second year after a burn that the *Spartina* marsh again becomes suitable for the dusky, this patchiness is extremely critical. In the early 1970s several wildfires burned portions of the St. Johns Refuge. The Recovery Team urged the Service to create fire lanes around and within the refuge to prevent fires from entering from ranchlands to the north, and to break up the refuge into management blocks so wild fires could more easily be controlled and fire could be used as a management tool for restoring suitable habitat. No action was taken and in the winter of 1975–76 two fires swept in from the north and burned over three-quarters of the refuge. After this fire the Service finally acquired the necessary equipment, constructed fire lanes, and conducted controlled burns. Although the habitat recovered after a few years, the dusky population did not.

The efforts of the U.S. Fish and Wildlife Service in behalf of the dusky seaside sparrow for the most part can be characterized as being too little and too late, nevertheless it alone does not bear the burden for the loss of this population. The financial constraints of recent years has forced the Service to assign lower priorities to endangered populations that are not full species. This is understandable, but no less unfortunate for the population heading for extinction. Priorities, of course, are not set on financial grounds alone. Public and political pressures also enter into these decisions. The dusky seaside sparrow is not a large, colorful or dramatic species easily seen by the public, such as is the whooping crane, California condor, or peregrine falcon. Most of the public in Brevard County have never heard of the dusky, let alone seen it. Its relegation to subspecies status, just as recovery efforts were being undertaken, effectively removed a large segment of the birdwatching public from its constituency. Its extinction will not cause many ripples of concern.

HERBERT W. KALE, II

REFERENCES

Delany, M. F., W. P. Leenhouts, B. Sauselein, and H. W. Kale II. 1981. The 1980 Dusky Seaside Sparrow Survey. Florida Field Naturalist 9:64–67.

James, F. C. 1980. Miscegenation in the Dusky Seaside Sparrow BioScience 30:800–801.

Review of Bird Conservation Literature

Richard C. Banks, Helen S. Lapham, and L. Richard Mewaldt

THIS compilation of literature includes references to all aspects of the biology of species of birds on the list of endangered and threatened species published by the U.S. Department of the Interior and to many of the "blue list" species of *American Birds*. References on population declines and trends in certain other species and to reports of human-related activities that may have an effect on bird populations are also included. Most of the references are to works in the technical literature. To the extent possible, only articles containing original information are cited. Popular and semi-popular magazines and newsletters were not reviewed.

Following the citation for each publication, we have provided, in parentheses, the address of the article's senior author, if it was available. This will be useful if you wish to request a reprint.

Adler, E. T. 1980. Counting the Bald Eagle. The Florida Naturalist 53(4):8–9.
Aldrich, J. W. and R. W. Coffin. 1980. Breeding bird populations from forest to suburbia after thirty-seven years. Amer. Birds 34:3–7. (6324 Lakeview Drive, Falls Church, VA 22041).
Anderson, A. 1980. The effects of age and wear on color bands. J. Field Ornith. 51:213–219. (Culterty Field Station, Department of Zoology, Aberdeen University, Newburgh, Aberdeenshire, Scotland, AB4 0AA).
Anderson A. 1980. Band wear in the Fulmar. J. Field Ornith. 50:101–109. (Cul-

terty Field Station, Department of Zoology, Aberdeen University, Newburgh, Aberdeenshire, Scotland, AB4 0AA).

Anderson, D. R. and K. P. Burnham. 1980. Effect of delayed reporting of band recoveries on survival estimates. J. Field Ornith. 51:244–247. (Utah Cooperative Wildlife Research Unit, Utah State University, UMC 52, Logan, Utah 84322).

Andrews, S. 1981. Black-crowned Night Heron predation of Black-necked Stilt. 'Elepaio 41:86. (Brigham Young U.—Hawaii Campus, Mus. of Nat. Hist., Laie, Hawaii 96762).—A chick of a stilt, listed as endangered, was swallowed by a heron. Management of native predators is raised.

Anon. 1981. Regional briefs—Region 4. Endangered Species Tech. Bull. 6(2):2.— The 1980 Everglade Kite census yielded 651 birds, perhaps the most in Florida in the last 50 years.

Anon. 1980. Six Bald Eagles alive and well on Catalina Island. The Western Tanager 47(4):8.

Anon. 1980. Bald Eagle-Osprey survey report, U.S. Forest Service, Eastern Region, 1979. The Passenger Pigeon 42:73–76.—Nesting surveys on National Forests in Michigan, Wisconsin and Minnesota were carried on in four aerial flights. There were 132 successful nests in 269 eagle territories with 222 young fledging. 194 Osprey territories were occupied out of 304 though productivity is still low.

Arbib, R. 1979. The Blue List for 1980. Amer. Birds 33:830–835. (American Birds, 950 Third Avenue, New York, NY 10022).

Arendt, W. J., T. A. Vargas Mora, and J. W. Wiley. 1981. White-crowned Pigeon: status rangewide and in the Dominican Republic. Proc. Ann. Conf. S.E. Assoc. Fish and Wildl. Agencies 33:111–122. (Inst. Tropical Forestry, U.S. Forest Service, P.O. Box AQ, Rio Piedras, PR 00928).—*Columba leucocephala* populations have undergone a widespread decline as a result of habitat destruction and poor management. Human harassment was responsible for the failure of most breeding colonies inspected.

Baker, W. W., R. L. Thompson, and R. T. Engstrom. 1980. The distribution and status of Red-cockaded Woodpecker colonies in Florida: 1969–1978. Florida Field Naturalist 8:41–45. (Tall Timbers Research Station, Rt. 1, Box 160, Tallahassee, Florida 32312).

Bayer, R. 1979. Bald Eagle-Great Blue Heron interactions. Murrelet 60:31–33. (Department of Zoology, Oregon State University, Marine Science Center, Newport, Oregon 97365).

Beaver, D. L. 1980. Recovery of an American Robin population after earlier DDT use. J. Field Ornith. 51:220–228. (Department of Zoology, Michigan State University, East Lansing, Michigan 48824).

Beaver, D. L., R. G. Osborn, and T. W. Custer. 1980. Nest-site and colony characteristics of wading birds in selected Atlantic Coast colonies. Wilson Bull. 92:200–220. (U.S. Fish and Wildlife Service, Patuxent Wildlife Research Center, Laurel, Maryland 20811).

Bednarz, J. C., and J. J. Dinsmore. 1981. Status, habitat use, and management of Red-shouldered Hawks in Iowa. J. Wildl. Manage. 45:236–241. (Dept. Animal Ecology, Iowa State Univ., Ames, IA 50011).

Berger, A. J. 1980. Longevity of Hawaiian Honeycreepers in captivity. Wilson Bull. 92:263–264. (Department of Zoology, University of Hawaii, Honolulu, Hawaii 96822).

Binkley, C. S., and R. S. Miller. 1980. Survivorship of the Whooping Crane, *Grus americana*. Ecology 61:434–437. (School of Forestry and Environ. Studies, Yale Univ., New Haven, CT 06511).

Bisbee, R. 1980. Nesting Brown Pelicans at San Bernard NWR. The Spoonbill 29(5):10–11.—First fledged young in upper Texas coast in 20 years.

Bishop, R. C. 1980. Endangered species: an economic perspective. Trans. N. Am. Wildl. Nat. Resour. Conf. 45:208–218. (Dept. Agric. Economics, Univ. Wisc., Madison, WI 53706).

Bloom, P. H. 1980. The status of the Swainson's Hawk in California, 1979. Job Final Rep., Job II–8.0, Proj. W–54–R–12, 42 pp. (California Dept. Fish and Game, 1416 Ninth Street, Sacramento, CA 95814).

Blus, L. J., C. J. Henny, and T. A. Kaiser. 1980. Pollution ecology of breeding Great Blue Herons in the Columbia Basin, Oregon and Washington. Murrelet 61:63–71. (U.S. Fish and Wildlife Service, 480 S.W. Airport Road, Corvallis, Oregon 97330).

Boag, D. A., and V. Lewin. 1980. Effectiveness of three waterfowl deterrents on natural and polluted ponds. J. Wildl. Manage. 44:145–154. (Dept. Zool., Univ. Alberta, Edmonton T6G 2E9, Alta., Canada).—Human effigies mounted on floats were more effective than model falcons or moving reflectors and were effective in diminishing the number of waterfowl and shorebirds dying due to bitumen fouling in a polluted tailings pond.

Bonney, R. E. Jr. 1979. Wintering Peregrine Falcon populations in the eastern United States, 1940–1975: A Christmas Bird Count Analysis. Amer. Birds 33:695–697. (Rte 2, Mohawk Trail, Charlemont, MA 01339).

Boyce, D. A. Jr. 1980. California Peregrine Falcon reproductive success and protective effort in 1980. 19 pp. (U.S. Fish and Wildlife Service, 2800 Cottage Way, Sacramento, CA 95825).

Beuch, R. R. 1980. Vegetation of a Kirtland's Warbler breeding area and 10 nest sites. The Jack-Pine Warbler 58:59–72. (North Central Forest Experiment Station, 1992 Folwell Ave., St. Paul, Minn. 55108).—Differences in tree density and age, understory vegetation and openings were studied at two breeding sites of Kirtland's Warblers (*Dendroica kirtlandii*). Though there were indications of importance of different variables, more studies covering wider range of jack pine ages are needed to determine exact needs for better management of the species.

Burger, J. and J. Shisler. 1980. Colony and nest site selection in Laughing Gulls in response to tidal flooding. Condor 82:251–258. (Department of Biology, Livingston College, Rutgers University, New Brunswick, New Jersey 08903.)

Burtt, E. H. Jr. 1979. Wild Mallard stocking in a large marsh habitat. Can. Field-Nat. 93:55–62.

Butler, R. G. and P. Lukasiewicz. 1979. A field study of the effect of crude oil on Herring Gull (*Larus argentatus*) chick growth. Auk 96:809–812. (The Mount Desert Island Biological Laboratory, Salsbury Cove, Maine 04672).

Byrd, G. V., and D. Moriarty 1980. Treated chicken eggs reduce predation on Shearwater eggs. 'Elepaio 41:13–15. (Aladdin Route Box 160E, Colville, WA 99114).—The Common Myna (*Acridotheres tristis*) has been a serious predator on eggs of the Wedge-tailed Shearwater (*Puffinus pacificus*). By placing bad-tasting chicken eggs near burrows at the beginning of laying, Mynas were conditioned against taking eggs.

Byrd, G. V., G. J. Divoky, and E. P. Bailey. 1980. Changes in marine bird and mammal populations on an active volcano in Alaska. Murrelet 61:50–62. (Hawaiian Islands National Wildlife Refuge, Box 87, Kilauea, Hawaii 96754).

Cairns, W. E., and I. A. McLaren. 1980. Status of the Piping Plover on the East Coast of North America. Amer. Birds 34:206–208. (Biology Department, Dalhousie University, Halifax, Nova Scotia B3H 4J1).

Calef, G. W., and D. C. Heard. 1979. Reproductive success of Peregrine Falcons and other raptors at Wager Bay and Melville Peninsula Northwest Territories. Auk 96:662–674. (Wildlife Service, Government of the Northwest Territories, Yellowknife, Northwest Territories, X1A 2L9 Canada).

Chabreck, R. H. 1980. Effects of marsh impoundments on coastal fish and wildlife resources. pp. 1–6. *In* Proc. Gulf of Mexico Coastal Ecosystem Workshop. U.S. Fish Wildl. Svc., Biol. Svcs. Prog. (FWS/OBS-80/30). (Available from NCET, NASA/Slidell Computer Complex, 1010 Gause Blvd., Slidell, LA 70458).

Conant, S. 1981. Recent observations of endangered birds in Hawaii's National Parks. 'Elepaio 41:55–61. (Dept. of General Science, Univ. of Hawaii at Manoa, 2450 Campus Rd., Honolulu, HI 96822).—Management plans are included.

Crawford, J. A., and D. K. Edwards. 1980. Winter grazing preferences of geese on a dredged-material island. Murrelet 61:106–108. (Department of Fisheries and Wildlife, Oregon State University, Corvallis, Oregon 97331).

Crawford, R. L. 1980. Wind direction and the species composition of autumn TV tower kills in northeastern Florida. Auk 97:892–895. (Tall Timbers Research Station, Route 1, Box 160, Tallahassee, Florida 32312).

Custer, T. W., R. G. Osborn, and W. F. Stout. 1980. Distribution, species abundance, and nesting-site use of Atlantic Coast colonies of Herons and their allies. Auk 97:591–600. (U.S. Fish and Wildlife Service, Patuxent Wildlife Research Center, Laurel, Maryland 20811).

Davis, T. E. and G. J. Niemi. 1980. Larid breeding populations in the western tip of Lake Superior. The Loon 52:3–14. (Dept. of Bio., U. of Minn. Duluth, Duluth, MN 55812).—Common Terns, which along with Ring-billed and Herring Gulls make up the Larid population, are 63% of total but are in competition with ring-bills and disturbed by humans. Main colony in a development area so alternate nesting site is needed.

Degange, A. R., and R. C. Newby. 1980. Mortality of sea birds and fish in a lost salmon driftnet. Mar. Poll. Bull. 11:322–323. (U.S. Fish and Wildlife Service, 791 8th St., Arcata, CA 95521).—99 dead seabirds found in 1500 m of floating net; mostly *Puffinus*.

Dekker, D. 1979. Characteristics of Peregrine Falcons migrating through central Alberta, 1969–1978. Can. Field-Nat., 93:296–302.

Desante, D. 1980. So gentle a ghoul. PRBO Newsletter No. 51. (Point Reyes Bird Observatory, Stinson Beach, CA 94970).

Devlin, W. J., J. A. Mosher, and G. J. Taylor. 1980. History and present status of the Red-migratory behavior of warblers. J. Field Ornith. 51:254–269. (Long Point Bird Observatory, P.O. Box 160, Port Rowan, Ontario, Canada N0E 1M0).—Differential responses of flight to bodies of water and to night electric lights are assessed.

Eckert, K. R. 1980. The Arctic Loon in Minnesota: a revised look at its status and identification. Loon 52:59–61. (9735 North Shore Drive, Duluth, MN 55804).

Erskin, A. J. 1980. Man's influence on potential nesting sites and populations of swallows in Canada. Can. Field-Nat. 93:371–377.

Feldman, E. S. 1980. Two baby eagles hatched. The Conservationist 35:38.—Two Bald Eagles hacked out into the wild have hatched out young in New York State. The Bald Eagle Restoration Program is part of the State's Division of Fish and Wildlife.

Ferrell, B. R., et al. 1980. Analysis of a Kentucky blackbird-starling roost population from birds killed by PA-14 treatment. The Kentucky Warbler 56:72–77. (Dept. of Biology, Western Kentucky Univ., Bowling Green, KY 42101).—Under a random sampling technique rather than a casual one, per cent of mortality of species at roost reflected per cent using roost. Technique allows relatively accurate analysis of roost species and sex composition and perhaps other parameters as well, from the mortality by use of PA-14.

Fitch, John H. 1980. The need for comprehensive wildlife programs in the United States: a summary. Council on Environmental Quality, 722 Jackson Place, N.W., Washington, D.C. 20006. 23 pp.

Flickinger, E. L., K. A. King, W. A. Stout, and M. M. Mohn. 1980. Wildlife hazards from Furadan 3G applications to rice in Texas. J. Wildl. Manage. 44:190–197. (Patuxent Wildl. Res. Cntr., Gulf Coast Field Sta., P.O. Box 2506, Victoria, TX 77901).—Furadan (carbofuran) causes less avian mortality than aldrin but does kill some birds.

Forsman, E. D. 1980. Habitat utilization by Spotted Owls in the west-central Cascades of Oregon. PhD Thesis, Oregon State University, Corvallis, Oregon. 95 pp. (Department of Fish and Wildlife, 104 Nash Hall, Oregon State University, Corvallis, OR 97331).

Fox, G. A., and T. Donald. 1980. Organochlorine pollutants, nest-defense behavior and reproductive success in Merlins. Condor 82:81–84. (Wildlife Toxicology Division, Canadian Wildlife Service, National Wilslife Research Centre, Department of the Environment, Ottawa, Ontario, Canada K1A 0E7).

Gagne, W. C. 1981. Hawai'i Audubon Society stops bulldozing in Kaua'i forest. 'Elepaio 41:92–93. (c/o Hawaii Audubon Soc., P.O. Box 22832, Honolulu, Hawaii 96822).—200 acres of primarily native Koa forest, home of endemic birds, bulldozed by State without regard to state environmental laws.

Gill, R., Jr., and L. R. Mewaldt. 1979. Dispersal and migratory patterns of San Francisco Bay produced herons, egrets, and terns. N. A. Bird Bander 4:4–13. (U.S. Fish and Wildlife Service, 1011 E. Tudor Road, Anchorage, Alaska 99503).

Gochfeld, M. 1980. Tissue distribution of mercury in normal and abnormal young Common Terns. Mar. Poll. Bull. 11:362–366. (Occupational and Environmental Health, Dept. of Health, Box 1540, Trenton, NJ 08625).—Higher mercury levels in 4 terns with abnormal feather loss than in 12 normal terns.

Goldsberry, J. R., S. L. Rhoades, L. D. Schroeder, and M. M. Smith. 1980. Waterfowl status report, 1975. Special Scientific Report—Wildlife No. 226. U.S. Fish and Wildlife Service, Washington, DC. (see Larned et al.).

Goldwasser, S., D. Gaines, and S. R. Wilbur. 1980. The Least Bell's Vireo in California: a de facto endangered race. American Birds 34:742–745. (Department of Ecology and Evolution, University of Arizona, Tucson, AZ 85721).

Graber, R., and J. Graber, 1980. Comparison of bird kill and census. Illinois Aud. Bull. 192:20–23. (Illinois Nat. Hist. Surv., Urbana, IL 61801).—Comparison of censuses of forest edge with nearby TV tower kill.

Grier, J. W., J. M. Gerrard, G. D. Hamilton, and P. A. Gray. 1981. Aerial-visibility bias and survey techniques for nesting Bald Eagles in northwestern Ontario. J. Wildl. Manage. 45:83–92. (Zool. Dept., North Dakota State Univ., Fargo, ND 58105).

Griffin, C. R., J. M. Southern, and L. D. Frenzel. 1980. Origins and migratory movements of Bald Eagles wintering in Missouri. J. Field Ornith. 51:161–167. (Missouri Cooperative Wildlife Research Unit, 1123 Stephens Hall, University of Missouri, Columbia, Missouri 65211).

Harrington-Tweit, B. 1980. First records of the White-tailed Kite in Washington. Western Birds 11:151–153. (900 North Wilson, Olympia, Washington 98506).

Harrison, E. N., and L. F. Kiff. 1980. Apparent replacement clutch laid by wild California Condor. Condor 82:351–352. (Western Foundation of Vertebrate Zoology, 1100 Glendon Avenue, Los Angeles, California 90024).

Harvey, T. E. 1980. California Clapper Rail survey, 1978–1979. Job V–1.8 Endangered Wildlife Program, California Department of Fish and Game, 1416 Ninth St., Sacramento, CA 95814.

Henny, C. J. and J. A. Collins. 1980. Early concentrations of Brown Pelicans along southern Oregon coast. Murrelet 61:99–100. (U.S. Fish and Wildlife Service, Patuxent Wildlife Research Center, 480 S. W. Airport Road, Corvallis, Oregon 97330).

Henny, C. J., and T. E. Kaiser. 1979. Organochlorine and mercury residues in Swainson's Hawk eggs from the Pacific Northwest. Murrelet 60:2–5. (U.S. Fish and Wildlife Service, Pacific Northwest Field Station, Patuxent Wildlife Research Center, 480 S. W. Airport Road, Corvallis, Oregon 97330).

Hill, E. F., and V. M. Mendenhall. 1980. Secondary poisoning of Barn Owls with Famphur, an organophosphate insecticide. J. Wildl. Manage. 44:676–681. (U.S. FWS, Patuxent Wildl. Res. Cntr., Laurel, MD 20811).—Eating prey poisoned by famphur may cause secondary poisoning.

Howe, M. A. 1980. Problems with wing tags: evidence of harm to Willets. J. Field Ornith. 51:72–73. (Migratory Bird and Habitat Research Laboratory, Patuxent Wildlife Research Center, U.S. Fish and Wildlife Service, Laurel, Maryland 20811).

Hruska, K., and K. Aschim. 1980. Prince Albert, Saskatchewan, birdhouse trail. Blue Jay (Saskatchewan) 38:249–250. (841 22nd St. East, Prince Albert, Saskatchewan, Canada S6V 1N9).—Designed for bluebird nesting.

Hurst, E., M. Hehnke, and C. C. Goude. 1980. The destruction of riparian vegetation and its impact on the avian wildlife in the Sacramento River Valley, California. Amer. Birds 34:8–12. (2952 Sally Court, Santa Clara, California 95051).

Johnson, D. H. 1979. Estimating nest success: the Mayfield method and an alternative. Auk 96:651–661. (U.S. Fish and Wildlife Service, Northern Prairie Wildlife Research Center, Jamestown, North Dakota 58401).

Johnson, E. V. 1980. Recovery of California's first captive-produced, wild-fostered Peregrine Falcon. N.A. Bird Bander 5:14. (Biological Sciences Department, California Polytechnic State University, San Luis Obispo, CA 93407).

Kaiser, G. W., K. Fry, and J. G. Ireland. 1980. Ingestion of lead shot by Dunlin. Murrelet 61:37. (Canadian Wildlife Service, Delta, BC, Canada V4K 3Y3).

Kear, J., and A. J. Berger. 1980. The Hawaiian Goose, an experiment in conservation. Buteo Books, Vermillion, S.D. 154 pp. $30.—Review of the biology, decline, and propagation of the Nene, *Branta sandvicensis*, and efforts to restore the wild population in Hawaii. Ultimate success not yet known. Problems of propagation programs are discussed.

Keast, A., and E. S. Morton (Eds.). 1980. Migrant birds in the Neotropics: ecology, behavior, distribution, and conservation. Smithsonian Inst. Press, Washington, DC 20560. 576 pp.—40 papers presented at a symposium in October 1977.

Kessel, B. 1979. Avian habitat classification for Alaska. Murrelet 60:86–94. (University of Alaska Museum, Fairbanks, AK 99701).

Kiff, L. F. 1979. Bird egg collections in North America. Auk 96:746–755. (Western Foundation of Vertebrate Zoology, 1100 Glendon Avenue, Los Angeles, California 90024).

Kilham, L. 1980. Association of Great Egret and White Ibis. J. Field Ornith. 51:73–74. (Department of Microbiology, Dartmouth Medical School, Hanover, NH 03755).

King, K. A., S. Macko, P. L. Parker, and E. Payne. 1979. Resuspension of oil: probable cause of Brown Pelican fatality. Bull. Environ. Contam. Toxicol. 23:800–805. (Patuxent Wildl. Res. Cntr., Gulf Coast Field Sta., P.O. Box 2506, Victoria, TX 77901).

King, K. A., D. L. Meeker, and D. M. Swineford. 1980. White-faced Ibis populations and pollutants in Texas, 1969–1976. Southwestern Nat. 25:225–240. (Patuxent Wildl. Res. Cntr., Gulf Coast Field Sta., P.O. Box 2506, Victoria, TX 77901).—Populations of *Plegadis chihi* declined by 42%, related to eggshell thinning and reduced reproductive success.

Knight, R. L., J. B. Athearn, J. J. Brueggman, and A. W. Erickson. 1979. Observations on wintering Bald and Golden Eagles on the Columbia River, Washington. Murrelet 60:99–105. (Wildlife Science Group, College of Fisheries, University of Washington, Seattle, Washington 98185).

Kress, S. W., ed. 1980. Arctic and Common Terns nest at Eastern Egg Rock. Egg Rock Update 1980: 1.—Gull Control and visual and sound models result in first terns nesting on E.Egg Rock, Maine in 43 years.

Kress, S. W., ed. 1980. Transplanted Puffin breeds at Matinicus Rock. Egg Rock Update 1980: 3.—First known breeding of a transplanted Puffin though breeding in an established colony rather than nearby E. Egg Rock, Maine, its transplanted home.

Kuyt, E. 1979. Banding of juvenile Whooping Cranes on the breeding range in the Northwest Territories, Canada. N.A. Bird Bander 4:24–25. (Canadian Wildlife Service, 9942 - 108 Street, Edmonton, Alberta, T5K 2J5, Canada).

Larned, W. W., S. L. Rhoades, and K. D. Norman. 1980. Waterfowl status report, 1976. Special Scientific Report—Wildlife No. 227. U.S. Fish and Wildlife Service, Washington, DC. (Office of Migratory Bird Management, Laurel, MD 20811).—Two reports bound together (see Goldsberry et al., above). Results of midwinter waterfowl surveys, 1974–57 and 1975–76, waterfowl breeding population and production surveys, 1975 and 1976, and waterfowl harvest surveys, 1974–75 and 1975–76.

Larson, D. 1980. Increase in the White-tailed Kite populations of California and Texas—1944–1978. Amer. Birds 34:689–690. (4500 19th Street, Boulder, Colorado 80302).

Layman, S. A. 1980. Feeding and nesting behavior of the Yellow-billed Cuckoo in the Sacramento Valley. Job Final Rep., Job IV–1.42, Proj. E–W–3. 28 pp. (California Dept. Fish and Game, 1416 Ninth Street, Sacramento, CA 95814).

Leedy, D. L. 1979. An annotated bibliography on planning and management for urban-suburban wildlife. U.S. Fish and Wildlife Service, Biological Services Program, FWS/OBS–79/25. 256 pp. (Urban Wildlife Res. Center, Inc., 4500 Sheppard Lane, Ellicott City, MD 21043).—464 references., in 6 subject categories—urban environment, effects of urbanization, wildlife and environmental values, planning aspects, management, plants, and research and education; cross referenced; author index.

Leedy, D. L., L. E. Dove, and T. M. Franklin. 1980. Compatibility of fish, wildlife, and floral resources with electric power facilities and lands: an industry survey analysis. Edison Electric Inst., 1111 19th St. NW, Washington, DC 20036. 130 pp.—"The extent of use by fish and wildlife and the occurrence of unique, threatened, or endangered species on electric utility lands is impressive."

Lehman, R. N. 1979. A survey of selected habitat features of 95 Bald Eagle nest sites in California. Job Final Rep., Job V–1.51, Proj. E–W–2. 23 pp. (California Dept. Fish and Game, 1416 Ninth Street, Sacramento, CA 95814).

Levenson, H., and J. W. Bee. 1980. Bald Eagle use of Kansas river riparian habitat in northeastern Kansas. Kans. Ornith. Soc. Bull. 31:28–37. (Mus. Nat. Hist. and Dept. Systematics and Ecol., U. Kansas, Lawrence, KS 66045).— The riparian habitat is excellent for wintering Eagles. (*Haliaeetus leucocephalus*). Favorable and unfavorable conditions currently in habitat discussed and management needs listed.

Lingle, G. R., and N. F. Sloan. 1980. Food habits of White Pelicans during 1976 and 1977 at Chase Lake National Wildlife Refuge, North Dakota. Wilson Bull. 92:123–125. (Department of Forestry, Mich. Tech, Univ., Houghton, Michigan 49931).

Locke, B.A., R. N. Conner, and J. C. Kroll. 1979. Red-cockaded Woodpecker stuck in cavity entrance resin. Bird-Banding 50:368–369. (School of Forestry, Stephen F. Austin State University, Nacogdoches, Texas 75962).

Lowe, R. L. 1980. Bald Eagles restoration efforts at LBL. Kentucky Warbler 56:71. (Land Between the Lakes, Tenn. Valley Auth., Golden Pond, KY 42231).— Two young Bald Eagles (*Haliaeetus leucocephalus*) from Wisconsin were hacked out successfully from Land Between Lakes, beginning an effort to restore eagle nesting to western Tennessee and Kentucky.

Lund, T. A. 1980. American wildlife law. Univ. of Calif. Press, Berkeley, CA. 171 pp. $10.95.

MacCarter, D. L. and D.S. MacCarter. 1979. Ten-year nesting status of Ospreys at Flathead Lake, Montana. Murrelet 60:24–49. (DLM, 20856 Isle Avenue West, Lakeville, Minnesota 55044).

Macmillen, R. E., and F. L. Carpenter. 1980. Evening roosting flights of the Honeycreepers (*Himatione sanguinea* and *Vestiaria coccinea*) on Hawaii. Auk 97:28–37. (Department of Ecology and Evolutionary Biology, University of California, Irving, California 92717).

Marcot, B. G., and J. Gardetto. 1980. Status of the Spotted Owl in Six Rivers National Forest, California. Western Birds 11:79–87. (Fisheries and Wildlife Branch, Six Rivers National Forest, 507 F. Street, Eureka, California 95501).

Maroldo, G. K. 1980. Crip: The constant dancer. Blue Jay (Saskatchewan) 38:147–161. (Texas Lutheran College, Sequin, Texas 78155).—Review of the life of "Crip," the captive Whooping Crane (*Grus americana*), and all his progeny. Also discussion of artificial factors affecting his life history.

Marsden, J. E., T. C. Williams, V. Krauthamer, and H. Krauthamer. 1980. Effect of nuclear power plant lights on migrants. J. Field Ornith. 51:315–318. (Dept. Neurobiology and Behavior, Cornell University, Ithaca, NY 14850).

Martin, T. E. 1980. Diversity and abundance of spring migratory birds using habitat islands on the Great Plains. Condor 82:430–439. (Department of Ecology, Ethology and Evolution, University of Illinois, Vivarium Building, Champaign, Illinois 61820).

McElroy, T. and G. Divoky. 1979. Seabirds, Eskimos and ice: Cooper Island '79. PRBO Newsletter No. 49. (Point Reyes Bird Observatory, Stinson Beach, CA 94970).

McEwan, L. C., and D. H. Hirth. 1980. Food habits of the Bald Eagle in northcentral Florida. Condor 82:229–231. (School of Forest Resources and Conservation, University of Florida, Gainesville, Florida 32611).

McKelvey, R. W., and D. W. Smith. 1979. A black bear in a Bald Eagle nest. Murrelet 60: 106–107. (Canadian Wildlife Service, Box 340, Delta, BC V4K 3Y3).

McKinley, D. 1980. The balance of decimating factors and recruitment in extinction of the Carolina Parakeet. The Indiana Aud. Quart. 58:8–18, 50–61, 103–114. (Dept. of Biological Sciences, State Univ. of NY., Albany, NY 12222).

Minsky, D. 1980. Preventing fox predation at a Least Tern colony with an electric fence. Jour. Field Ornith. 51:180–181. (Cape Cod National Seashore, South Wellfleet, MA 02663).

Morris, R. D., I. R. Kirkham, and J. W. Chardine. 1980. Management of a declining Common Tern colony. J. Wildl. Manage. 44:241–245. (Dept. of Psychology, Memorial Univ. of Newfoundland, St. John's, Newfoundland A1B 3X9). Control of vegetation growth and prevention of gull nesting increased tern reproductive success on Gull Island, Lake Ontario.

Nelson, L. K., D. R. Anderson, and K. P. Burnham. 1980. The effect of band loss on estimates of annual survival. J. Field Ornith. 51:30–38. (Utah State University, Department of Wildlife Science, Logan, Utah 84322).

Newman, J. R. 1979. Effects of industrial air pollution on wildlife. Biol. Conserv. 15:181–190.

Nichols, J. D. and C. M. Haramis. 1980. Sex-specific differences in winter distribution patterns of Canvasbacks. Condor 82:406–416. (Migratory Bird and Habitat Research Lab., U.S. Fish and Wildlife Service, Laurel, Maryland 20811).

Nilsson, G. 1981. The bird business, a study of the commercial cage bird trade. 2nd ed. 121 pp. Animal Welfare Inst., P.O. Box 3650, Washington, DC 20007. $5.00. A study of the cage bird trade, primarily into the U.S., with chapters on established exotics and threatened species. Includes data on birds imported into U.S. 1977–80.

Ogden, J. C., and S. A. Nesbitt. 1979. Recent Wood Stork population trends in the United States. Wilson Bull. 91:512–523. (National Audubon Society Research Department, 115 Indian Mound Trail, Tavernier, Florida 33070).

Olson, J. F. 1980. Osprey of the Turtle-Flambeau Flowage. The Passenger Pigeon 42:101–102. (Box R, Mercer, WI 54547).—Area has had the largest concentration of a nesting population in NW Wisconsin. Since the banning of DDT, young produced have increased. Each site has a management plan to minimize human disturbance. Artificial nest platforms have been built and accepted.

Parker, J. W., and J. C. Ogden. 1979. The recent history and status of the Mississippi Kite. Amer. Birds 33:119–129. (Department of Biology and Environmental Resources Center, State University College, Fredonia, NY 14063).

Perry, M. C., F. Ferrigno, and F. H. Settle. 1980. Rehabilitation of birds oiled on two mid-Atlantic estuaries. Proc. Ann. Conf. S.E. Assoc. Fish and Wildl. Agencies 32:318-325. (U.S. FWS, Migratory Bird and Habitat Res. Lab., Laurel, MD 20811).—Estimates 52,500 birds died as a result of 7 major oil spills on Delaware River and Chesapeake Bay. Ruddy Duck, Horned Grebes, and Oldsquaws were heavily affected. 82% of oiled waterfowl captured alive died during or shortly after cleaning.

Pfannmuller, L. A. and D. G. Wells. 1981. Minnesota natural heritage program, breeding bird elements. The Loon 53:5–8. (Minn. Natural Heritage Prog., Rsh. and Policy Sect., Minnesota Dept. of Nat. Res., Box 11, Centennial Office Bdg., St. Paul, MN 55155).—Based on a study initiated by Minn. Dept. of Nat. Resources in 1979, birds needing special attention are listed according to status.

Pfeiffer, M. B. 1980. The herons of Shooters Island. The Conservationist 34(6):EQN II–EQN VI.—Shooters Island off New York City, is the only local colony of mixed herons. Seven species and some 200 pairs breed. Business interests want to remove the island to clear the channel and conservationists are fighting it.

Platt, S. W. 1980. Longevity of herculite leg jess color markers on the Prairie Falcon (*Falco mexicanus*). J. Field Ornith. 51:281–282. (Department of Zoology, Brigham Young University, Provo, Utah 84602).

Pratt, H. M. 1980. Directions and timing of Great Blue Heron foraging flights from a California colony: implications for social facilitation of food finding. Wilson Bull. 92:489–496. (Point Reyes Bird Observatory, 4990 Shoreline Highway, Stinson Beach, California 94970).

Primack, M. 1980. ORVs in our national seashores. Nat. Parks and Conserv. Mag. 54(11):4–7.—In past 20 years, ORV registration has grown 2000%. A 5-year study on Cape Cod National seashore, which allows ORVs, determined it had "no carrying capacity for vehicular impacts on coastal ecosystems." In Cape Cod, one ORV driver went around barricades protecting Tern colonies and killed three young including 2 Arctic young representing 50% of the nests on the Cape.

Pritchard, P. C. H. 1981. Captive breeding. The Florida Naturalist 54(1):9–9, 11, 15.

Pruett-Jones, S. G., M. A. Pruett-Jones, and R. L. Knight. 1980. The White-tailed Kite in North and Middle America: current status and recent population changes. Amer. Birds 34:682–688. (Museum of Vertebrate Zoology, 2593 LSB, Berkeley, CA 94720).

Raynor, G. S. 1980. Bobwhite population changes on Long Island, New York 1959–1978. Amer. Birds 34:691–694. (Schultz Road, Manorville, NY 11949).

Robertson, R. J., and N. J. Flood. 1980. Effects of recreational use of shorelines on breeding bird populations. Can. Field-Natur., 94:131–138. (Dept. Biol., Queen's Univ., Kingston, Ont. K7L 3N6).

Rosahn, J. W. 1980. Bobwhite population—a 10-year study. Amer. Birds 34:695–697. (206 Ellwood Road, Kensington, Conn. 06037).

Ryel, L. A. 1980. Kirtland's Warbler status, June, 1979. The Jack-Pine Warbler 58:30–32. (Office of Surveys and Statistics, Mich. Dept. of Natural Resources, Box 30028, Lansing, Mich. 48909).

Ryel, L. A. 1980. Results of the 1980 census of Kirtland's Warblers. The Jack-Pine Warbler 58:142–145. (Wildlife Division, Mich. Dept. of Natural Resources, Box 30028, Lansing, Mich. 48909).

Sandilands, A. P. 1980. Artificial nesting structures for Great Blue Herons. Blue Jay (Saskatchewan) 38:186–188. (P.O. Box 146, Plattsville, Ontario, Canada N0J 1S0).

Schreiber, R. W. 1979. Reproductive performance of the Eastern Brown Pelican, *Pelecanus occidentalis*. Contributions in Science No. 317, Nat. Hist. Mus. of Los Angeles County: 1–43. (Natural History Museum of Los Angeles County, 900 Exposition Blvd., Los Angeles, CA 90007).

Schreiber, R. W. 1980. Nesting chronology of the Eastern Brown Pelican. Auk

97:491–508. (Natural History Museum, Los Angeles County, 900 Exposition Boulevard, Los Angeles, California 90007).

Sherrill, D. M., and V. M. Case. 1980. Winter home ranges of 4 clans of Red-cockaded Woodpeckers in the Carolina Sandhills. Wilson Bull. 92:369–375. (Department of Biology, Davidson College, Davidson, North Carolina 28036).

Siderits, K. 1980. 1980 Bald Eagle, Osprey and Blue Heron nesting, Superior National Forest. The Loon 52:193. (Wildlife Biologist, Superior National Forest, P.O. Box 338, Duluth, MN 55801).

Siderits, K. 1980. 1980 Bald Eagle, Osprey and Blue Heron nesting, Superior National Forest. The Loon 52:193. (Wildlife Biologist, Superiof Nat. Forest, P.O. Box 338, Duluth, MN 55801).—The good year resulted in 39 young eagles (1.1/nest), 52 Osprey (1.4/nest) and 14 heron colonies with 9 to more than 150 nests each.

Siegel-Causey, D. 1980. Progenicide in Dougle-crested Cormorants. Condor 82:101. (Department of Ecology and Evolutionary Biology, University of Arizona, Tucson, Arizona 85721).

Sincock, J. L. and J. M. Scott. 1980. Cavity nesting of the Akepa on the island of Hawaii. Wilson Bull. 92:261–263. (Patuxent Wildlife Research Center, U.S. Fish and Wildlife Service, P.O. Box 197, Koloa, Hawaii 96756).

Spear, L. 1980. Band loss from the Western Gull on Southeast Farallon Island. J. Field Ornith. 51:319–328. (Point Reyes Bird Observatory, 4990 Shoreline Highway, Stinson Beach, CA 94970).

Spitzer, P. and A. Poole. 1980. Coastal Ospreys between New York City and Boston: a decade of reproductive recovery 1969–1979. American Birds 34:234–241. (Section of Ecology and Systematics, Langmuir Laboratory, Cornell University, Ithaca, NY 14850).

Stamm, A. L. and J. Durell. 1980. The 1980 Bald Eagle count in Kentucky. The Kentucky Warbler 56:55–58. (Dept. of Fish and Wildlife, Frankfort, Kentucky 40601).

Stapleton, J. and E. Kiviat. 1979. Rights of birds and rights of way. (Vegetation management on a railroad causeway and its effect on breeding birds). Amer. Birds 33:7–10. (John Burroughs Nature Sanctuary, P.O. Box 220, West Park, NY 12493).

Stauffer, D. F., and L. B. Best. 1980. Habitat selection by birds of riparian communities: evaluating effects of habitat alterations. J. Wildl. Manage. 44:1–15. (Dept. Anim. Ecology, Iowa State Univ., Ames IA 50011).—Management for maximum diversity can be detrimental to rare species.

Steenhof, K., S. S. Berlinger, and L. H. Fredrickson. 1980. Habitat use by wintering Bald Eagles in South Dakota. J. Wildl. Manage. 44:798–805. (Gaylord Mem. Lab., Univ. Missouri, Puxico, MO 63960).—Distribution of Bald Eagles at a winter site apparently is influenced most by location of food and areas protected from wind.

Stendell, R. C. 1980. Dietary exposure of Kestrels to lead. J. Wildl. Manage. 44:527–530. (U.S. FWS, Northern Prairie Wildl. Res. Cntr., Jamestown, ND 58401).—Survival of Kestrels fed lead shot under experimental conditions was not affected, although liver residues were increased.

Stendell, R. C., J. W. Artmann, and E. Martin. 1980. Lead residues in Sora Rails from Maryland. J. Wildl. Manage. 44:525–527. (Northern Prairie Wildlife Res. Cntr., Jamestown, ND 58401).—7.4% of the Soras studied had ingested lead shot. Residues in some livers were within the range considered to be lethal in waterfowl.

Stenzel, L. 1980. Feast and famine in the Gulf of Farallones PRBO Newsletter No. 50. (Point Reyes Bird Observatory, Stinson Beach, CA 94970).

Stewart, P. A. 1980. Population trends of Barn Owls in North America. Amer. Birds 34:698–700. (203 Mooreland Drive, Oxford, NC 26565).

Sykes, P. W. Jr. 1979. Status of the Everglade Kite in Florida. Wilson Bull. 91:495–511. (U.S. Fish and Wildlife Service, Patuxent Wildlife Research Center, Field Station, P.O. Box 2077, Delray Beach, Florida 33444).

Sykes, P. W. Jr. 1980. Decline and disappearance of the Dusky Seaside Sparrow from Merritt Island, Florida. Amer. Birds 34:728–737. (U.S. Fish and Wildlife Service, Patuxent Wildlife Research Center, Field Station, P.O. Box 2077, Delray Beach, Florida 33444).

Thill, J. F. 1980. The Month. The Prothonotary 46:117–118. (36 Allegany Ave., Kenmore, NY 14217).—Second greatest 'kill' at towers south of Buffalo since counting began in 1967. 722 birds of 36 species on Sept. 18. The 20th and 21st totalled some 600 of 32 species.

Tilghman, N. G. 1980. The Black Tern Survey, 1979. The Passenger Pigeon 42:1–8.—Terns decreasing in Wisconsin and indications they are decreasing countrywide. Wisconsin Dept. of Natural Resources has initiated a thorough wetlands survey using two different methods. These data will give a base line to detect significant changes in distribution and abundance in the future.

Tomlinson, W. H., Jr. and F. L. Haines III. 1980. Use of Bluebird nest boxes in Coastal South Carolina. The Chat 44:70–75. (Anderson-Tully Co., Vicksburg, Miss. 39180).

Van Balen, J. B., and P. D. Doerr. 1980. The relationship of understory vegetation to Red-cockaded Woodpecker activity. Proc. Ann. Conf. S.E. Assoc. Fish and Wildl. Agencies 32:82–92. (Dept. Zool., North Carolina State Univ., Raleigh, NC 27650).—Preference for open areas.

Van Daele, L. J. and H. A. Van Daele. 1980. Observations of breeding Bald Eagles in Idaho. Murrelet 61:108–110. (Department of Biological Sciences—Zoology, University of Idaho, Moscow, Idaho 83843).

Wahl, T. R. and D. Heinemann. 1979. Seabirds and fishing vessels: Co-occurrence and attraction. Condor 81:390–396. (Department of Biology, Western Washington University, Bellingham, Washington 98225).

Wallace, M. P., P. G. Parker, and S. A. Temple. 1980. An evaluation of patagial markers for cathartid vultures. J. Field Ornith. 51:309–314. (Department of Wildlife Ecology, University of Wisconsin, Madison, WI 53706).

Ward, L. D. and J. Burger. 1980. Survival of Herring Gull and domestic chicken embryos after simulated flooding. Condor 82:142–148. (Department of Zoology, Rutgers University, New Brunswick, New Jersey 08903).

Weller, M. W. 1980. The island waterfowl. Iowa State Univ. Press, Ames, IA. 121 pp. $10.95.—a preliminary report. Amer. Birds 33:252. (U.S. Fish and

Wildlife Service, Patuxent Research Center, California Field Station, Ojai, California 93023).

Wilbur, S. R. 1980. Estimating the size and trend of the California Condor population, 1965–1978. Calif. Fish and Game 66:40–48. (U.S. Fish and Wildlife Service, Patuxent Wildlife Research Center, Ojai, CA 93023).

Wilbur, S. R., and L. F. Kiff. 1980. The California Condor in Baja California, Mexico. Amer. Birds 34:856–859. (U.S. Fish and Wildlife Service, 1190 E. Ojai Avenue, Ojai, CA 93023).

Wilbur, S. R., P. D. Jorgensen, B. W. Massey, and V. A. Basham. 1979. The Light-footed Clapper Rail: an update. Amer. Birds 33:251. (U.S. Fish and Wildlife Service, 1190 E. Ojai Avenue, Ojai, California 93023).

Wilbur, S. R., R. D. Mallette, and J. C. Borneman. 1979. California Condor survey, 1978. Calif. Fish and Game 65:183–184. (U.S. Fish and Wildlife Service, 1190 E. Ojai Avenue, Ojai, California 93023).

Wilkinson, G. S. and K. R. Debban. 1980. Habitat preferences of wintering diurnal raptors in the Sacramento Valley. Western Birds 11:25–34. (Department of Biology, C–016, University of California at San Diego, La Jolla, California 92093).

Wilson, R. A. 1980. Snowy Plover nesting ecology on the Oregon Coast. MA Thesis, Oregon State University, Corvallis, Oregon. 41 pp. (Department of Fish and Wildlife, Oregon State University, Corvallis, OR 97331).

Winter, J. 1980. Status and distribution of the Great Gray Owl in California. Job Final Report. Job II-9, Proj. W–54–R–12. 37 pp. (California Dept. Fish and Game, 1416 Ninth Street, Sacramento, CA 95814).

Yparraguirre, D. R. 1979. Distribution, migration, and mortality of Aleutian Canada Geese in California, 1978–1979. Job Final Report, Job V–1.41, Proj. E–W–3. 30 pp. (California Dept. of Fish and Game, 1416 Ninth Street, Sacramento, CA 95814).

Contributors

GEORGE W. ARCHIBALD
International Crane Foundation
City View Road
Baraboo, Wisconsin 53913

RICHARD C. BANKS
National Wildlife Laboratory
National Museum of Natural History
Smithsonian Institution
Washington, D.C. 20560

JOHN H. BARCLAY
Peregrine Fund
Cornell Laboratory of Ornithology
159 Sapsucker Woods Road
Ithaca, New York 14850

TOM J. CADE
Peregrine Fund
Cornell Laboratory of Ornithology
159 Sapsucker Woods Road
Ithaca, New York 14850

JAMES B. ELDER
U.S. Fish and Wildlife Service
Fort Snelling
Twin Cities, Minnesota 55111

FRANCIS J. GRAMLICH
U.S. Fish and Wildlife Service
P.O. Box 800
Augusta, Maine 04330

NANCY F. GREEN
Bureau of Land Management
764 Horizon Drive
Grand Junction, Colorado 81501

JAMES W. GRIER
Zoology Department
North Dakota State University
Fargo, North Dakota 58105

HERBERT W. KALE II
Florida Audubon Society
1101 Audubon Way
Maitland, Florida 32751

WARREN B. KING
871 Dolley Madison Boulevard
McLean, Virginia 22101

JOEL V. KUSSMAN
National Park Service
Denver Service Center
Denver, Colorado 80225

HELEN S. LAPHAM
Cornell Laboratory of Ornithology
159 Sapsucker Woods Road
Ithaca, New York 14850

THOMAS E. LOVEJOY
World Wildlife Fund
1601 Connecticut Avenue, NW
Washington, D.C. 20009

JOHN E. MATHISEN
U.S. Forest Service
Chippewa National Forest
Cass Lake, Minnesota 56633

JAMES MATTSSON
U.S. Fish and Wildlife Service
Agassiz National Wildlife Refuge
Middle River, Minnesota 56737

L. RICHARD MEWALDT
Avian Biology Laboratory
San Jose State University
San Jose, California 95192

JOHN C. OGDEN
National Audubon Society
Condor Research Center
87 North Chestnut Street
Ventura, California 93001

JAY M. SHEPPARD
Office of Endangered Species
U.S. Fish and Wildlife Service
Washington, D.C. 20240

LESTER L. SHORT
American Museum of Natural History
Central Park West at 79th Street
New York, New York 10024

NOEL F. R. SNYDER
U.S. Fish and Wildlife Service
Condor Research Center
87 North Chestnut Street
Ventura, California 93001